# Theory and Practice of Weed Control

*Edited by*

**R.J. Stephens**

*School of Biological Science*
*University of Bath*

United Book Prints
*New Delhi*

Reprint: 2021

*Reprint in India*, 2014
By
United Book Prints
4806/24, Ansari Road, Darya Ganj,
New Delhi (India)
E-mail: unitedbookprints@gmail.com
Tel.: + 91 11 41511055

ISBN : 978-81-926860-7-3


Reprint in India at Hinglaj Offset

# CONTENTS

# PREFACE

Why another book about weeds? Authors often begin by justifying a further addition to those already available, but there is such a dearth of books on weed science for students that no justification is required. This book is neither a comprehensive textbook nor a weed control manual but instead an attempt to provide an overall view of weed biology and control, illustrating the principles with appropriate examples drawn from temperate and tropical horticultural and plantation crops.

Wherever possible the vernacular and scientific plant names in the text are those recommended in J. G. Dony, F. Perring and Catherine M. Rob (1974), *English Names of Wild Flowers*, published by Butterworth for the Botanical Society of the British Isles. Tropical weeds and other species that do not appear in that list have been named according to the general practice in scientific journals and floras. Herbicides have been given the common names listed from time to time in *Weed Science* and *Weed Research*, and in those few cases where these lists do not agree the name given in *Weed Research* has been preferred. SI units of measurement are used throughout.

The references at the end of each chapter are intended to enable the reader to explore a particular interest more deeply. Most of those cited are modern and easily obtainable, but the older and historically interesting ones may only be obtainable from library loan services.

The author wishes to thank all those whose published works he has freely consulted during the preparation of this book. He also thanks the colleagues and friends who have read and commented upon parts of the book and especially Professor Leonard Broadbent who has patiently edited the whole book.

None of the information given in this book should be regarded as a recommendation, and anyone contemplating chemical weed control should follow directions given by the official advisory services in their region or by the manufacturers.

*Bath, 1981* R.J.S.

# PREFACE

# 1 EFFECTS OF WEED INFESTATION ON CROP YIELD AND QUALITY

## 1.1 WHAT ARE WEEDS?

Weeds are plants in the wrong place. For the purposes of this book any plant competing with cultivated plants or that in some other way interferes with man's legitimate activities is considered to be a weed. Algae, mosses, liverworts, ferns and flowering plants all can qualify as weeds. Annuals, rarely biennials and both herbaceous and woody perennials may all be weeds. Crop plants become weeds themselves when they persist or regenerate in succeeding crops where they may act as sources of pests and disease pathogens. Why control weeds? Early farmers probably harvested a mixture of several 'crops' including some species now regarded as weeds, but the unpleasant tasting, poisonous or least productive components of the mixture would soon have been actively discouraged. At first this was by hand pulling, then by hoeing, cultivation and crop rotation, and in very recent times by seed cleaning and the use of herbicides. But in many parts of the world the extent of cultivation is still determined by the area that each family can keep clean by hand during the critical early period of crop establishment.

## 1.2 LOSSES IN CROP PRODUCTION DUE TO WEEDS

All authorities agree that immense losses of food, fibre and other commodities are caused by pests, diseases and weeds both before and after harvest. These losses may reach 25 per cent in Europe, but in the less developed areas in Africa and Asia almost half of the potential food yield is lost (Altman and Campbell, 1977). Overall world losses due to weeds alone are considerable (Cramer, 1967; Chisaka, 1977). In addition to these direct losses caused by 'pests' of all kinds, there is an enormous loss of potential production for agronomic reasons such as inadequate nutrition and water supply and the use of poor quality or low yielding cultivars. Control of pests must therefore be viewed as an integral part of good husbandry.

Parker and Fryer (1974) suggest that world crop production can be divided into three categories, A, B and C, according to degree of sophistication in terms of adoption of modern methods such as use of pesticides, mechanisation and plant breeding. They estimate that in category A, the most highly developed agriculture, weeds are already so well controlled by herbicide and other means that losses average no more than 5 per cent of actual production and this makes significant improvement difficult. In the less advanced category B, where herbicides are used more sparingly, losses due to weeds probably average about 10 per cent and a reduction from 10 to 6 per cent would be a reasonable target to aim for. However, in the least developed category, C, in what could be called peasant agriculture, where herbicides and other modern aids are little used, average losses due to weeds exceed 25 per cent. Damage from pests and disease also tends to be greater in peasant farming but innovations and improvements are often delayed and discouraged by economic, social and political obstacles. Yet it is probably in peasant agriculture that such changes could have the greatest affect on crop production.

Crop loss data alone, however, fail to show the full cost of weeds to mankind. It is probable that more effort is devoted to hand weeding than to any other single human task, and under the hot and humid conditions of the tropics it is especially demanding. In peasant communities the daily necessity for children to work long hours in the fields effectively prevents any improvement in education. Invasion of rice paddy fields in Sri Lanka by Kariba Weed (*Salvinia molesta*) and of sugar cane plantations in India by perennial grass weeds that closely resemble the crop, both of which are difficult to control by any method, has led to formerly productive land being abandoned. In parts of Afghanistan seeds of a species of *Heliotropium* growing as a weed among wheat are harvested with the crop. Bread made from the contaminated flour contains a poisonous alkaloid which is thought to be responsible for a liver degeneration that has recently caused many thousands of deaths (Holm, 1976). The weeds Corncockle (*Agrostemma githago*) and Darnel (*Lolium temulentum*) caused similar, though less severe, problems in Europe until effective seed cleaning introduced in the nineteenth century was able to prevent seed of these species being sown with the crop. In countries where inadequately cleaned seed is planted, contamination of grain by Darnel seed still causes taint and at higher concentrations makes bread poisonous.

Wind pollinated weeds produce pollen which may cause hay fever in susceptible people. In North America the Ragweeds (mainly *Ambrosia artemesifolia*) release annually an estimated one million tonnes of pollen, and grasses rather more, together causing great human misery and disruption of production and services. Other species, including several species of *Rhus* native to North America and known collectively as Poison Ivies, can cause severe skin lesions when touched. Livestock also suffer pain and death from poisonous weeds, from barbed weed seeds that become embedded in eyes and other sensitive areas, and from weeds containing substances that cause extreme photosensitivity when eaten. Animals that eat these weeds suffer severe damage to unpigmented areas of skin when exposed to the sun (see Chapter 4).

Aquatic weeds, several species of which have spread rapidly during this century, can reduce substantially the flow of water in irrigation channels and drains, increase water loss due to evaporation, reduce fish numbers and make navigation of rivers difficult in addition to reducing yields of rice by direct plant competition (see Chapter 4).

## 1.3 WEED COMPETITION

### 1.3.1 Direct losses due to competition

Weeds cause direct losses by depriving crops of water, light and mineral nutrients. Quality may be impaired by the presence of weeds, often reducing the economic value more than reduced weight of crop would indicate. For example, weedy lettuce may fail to make firm, marketable heads (Roberts *et al.*, 1977), and growers may be penalised for offering peas for freezing containing unopened thistle flowers or green immature fruits of Black Nightshade (*Solanum nigrum*), both of which are similar in size and colour to peas.

The effects of weed competition in annual species are restricted to that year's crop. However, weed competition in perennial crops will often have repercussions in future years. For example, newly planted fruit trees and bushes are vulnerable to weed competition and stunted early growth inevitably reduces their future productivity (Chapters 10 and 11).

### 1.3.2 Biochemical competition

It is now established that plants can compete with one another at a biochemical level by means of substances capable of exerting an inhibitory effect on growth when vaporised from leaves, washed from leaves to the soil or exuded into the soil from roots (Harborne, 1977; Rice, 1974). Water extracts of plants commonly inhibit the growth of other species. In nutrient culture, liquid circulating between the roots of different species has been shown to have mutually inhibiting effects. There have been claims that decomposing rhizomes of some weeds, including Common Couch (*Agropyron repens*), produce inhibitory substances (Drost and Doll, 1980), a claim now supported by the demonstration that acetic acid produced by an overlying blanket of wet, rotting cereal straw inhibits wheat germination (Chapter 8).

False Flax (*Camelina alyssum*) is claimed to inhibit flax growth less by competition for water and nutrients than by direct inhibition of crop roots by phenolic substances washed from leaves into the soil by rain. *Salvia leucophylla*, a plant of mountainous areas of California, produces volatile terpenes, including camphor and camphene, that are leached into the soil where they inhibit plant growth for several metres from the *Salvia* plants. That substances exerting allelopathic effects are produced by living and dead plants under some circumstances is beyond doubt and the possibility that their action is sometimes important in crop situations cannot be ignored. But it is probable that most harmful effects of weeds on crops are the direct result of competition for the essentials for plant life, water, light and mineral nutrients, and that the claims for biochemical competition have been exaggerated and that it plays no more than a minor role in competition with crops (Welbank, 1963).

### 1.3.3 Weed competition and crop production: critical period of weed competition

A newly emerged crop seedling relying for food on that stored in the seed, would experience little competition from weed seedlings at a similar growth stage, if only because at that time total demands on water and minerals are small. Likewise a crop towards the end of its growth cycle, especially if it is senescing and ripening seed, will often be unharmed by the presence of weeds. Between these situations there should be a period when crops would be most susceptible and weed competition would be most severe. The existence of this period, known as the critical period for weed competition, was verified and quantified by Nieto *et al.* (1968) in Mexico, who demonstrated that maize was most vulnerable to weed competition between 10 and 30 days after emergence. Weeds present before and after the critical period, however, had a negligible effect on yield (see *Figure 1.1*). Their method was to weed several plots by hand allowing, at weekly intervals, some of the plots to become weedy and remain so until harvest. Other plots were left unweeded, but some of these, again at weekly intervals, were hand weeded and then kept clean. Their method has since been applied to many annual, biennial and perennial crops and Nieto's principle of a critical period has been found to be of general application (Scott *et al.*, 1979).

In onions, for example, in which germination and initial growth are slow, the crop competes poorly with annual weeds and when weed control fails low yields of small thick-necked bulbs are produced. Hewson and Roberts (1971 and 1973) found that annual weeds had little effect on the crop during the first 4–6 weeks after 50 per cent emergence, but that if weeds remained for another two weeks there was a 60 per cent reduction in yield; weeds that developed after six

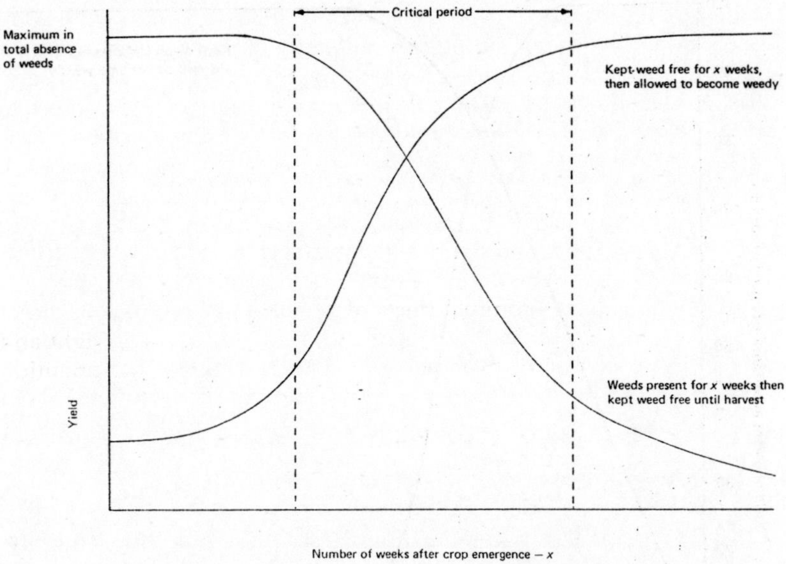


**Figure 1.1** The concept of critical period of weed competition, first proposed by Nieto, Brondo and Gonzalez, 1968. (Adapted from Bleasdale, 1973.)

weeks had little effect on yield (see *Figure 1.2*). Weed competition during the critical period, coinciding with the development of the third true leaf, led to yield reduction; continued competition beyond six or seven weeks resulted in death of up to half the bulbs. Why are annual weeds so detrimental to this crop? Annual weeds can achieve a greater maximum relative growth rate (about 180 mg/g/day) than onions (about 120 mg/g/day), and also achieve it sooner. This advantage over the crop has a marked effect on mineral uptake (see *Figure 1.3*), analysis showing that 7½-week-old weeds at 160–230/m$^2$ contained the equivalent of half of the nitrogen and one third of the potassium that had been applied as a base dressing, compared with the relatively small amount taken up by weedy onions. Late removal of weeds has little beneficial effect on yield because the period of vegetative growth which is essential for onion bulking has already passed.

Growers may be tempted to compensate for competition by weeds for plant nutrients by giving a top-dressing to a weakened crop (Lawson and Sieman, 1979), but generally the weeds benefit while the crop does not; Wild-oat (*Avena fatua*), Fat-hen (*Chenopodium album*), Charlock (*Sinapis arvensis*) and especially Common Chickweed

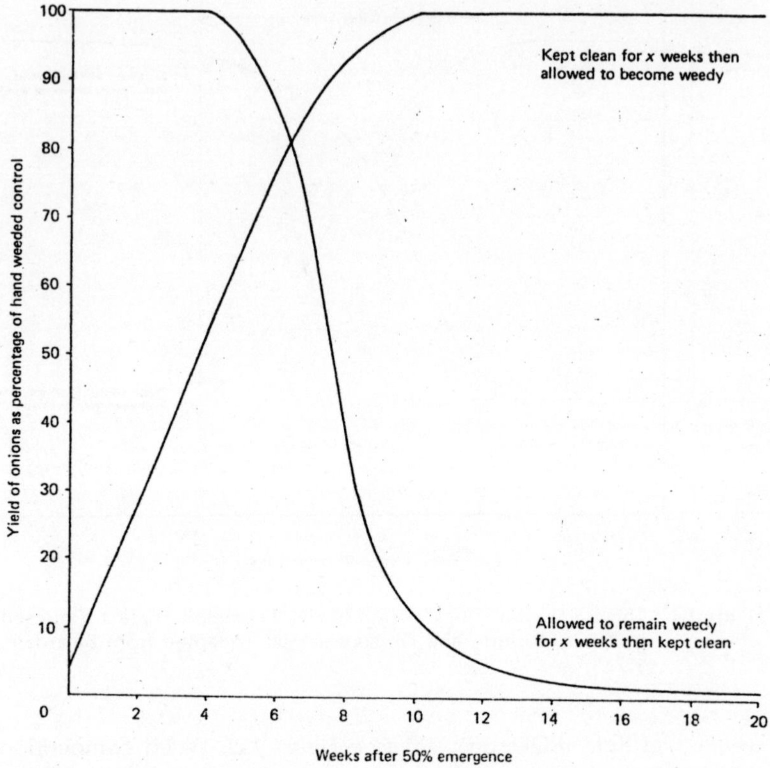


**Figure 1.2** Onion yields expressed as a percentage of hand weeded control when allowed to remain weedy for different periods. (From Hewson and Roberts, 1971.)

(*Stellaria media*) have all been shown to be very effective at exploiting added nitrogen.

Although yields are seldom reduced by the continued presence of weeds once the critical period for weed competition has passed, weeds may nevertheless be a serious handicap to harvesting. Patches of green weeds force combine harvesters to work more slowly and weeds increase the wastage of grain. A particular problem in Australia is caused by the wiry flowering stems of Skeleton Weed (*Chondrilla juncea*), an alien from Southern Europe, which can make wheat harvesting virtually impossible. Some weeds that cause skin irritation, especially Small Nettle (*Urtica urens*), create difficulties when picking strawberries, raspberries and blackcurrants.

The effects of weed competition on perennial plants are usually more complex (Chapters 10 and 11). When raspberries are planted in

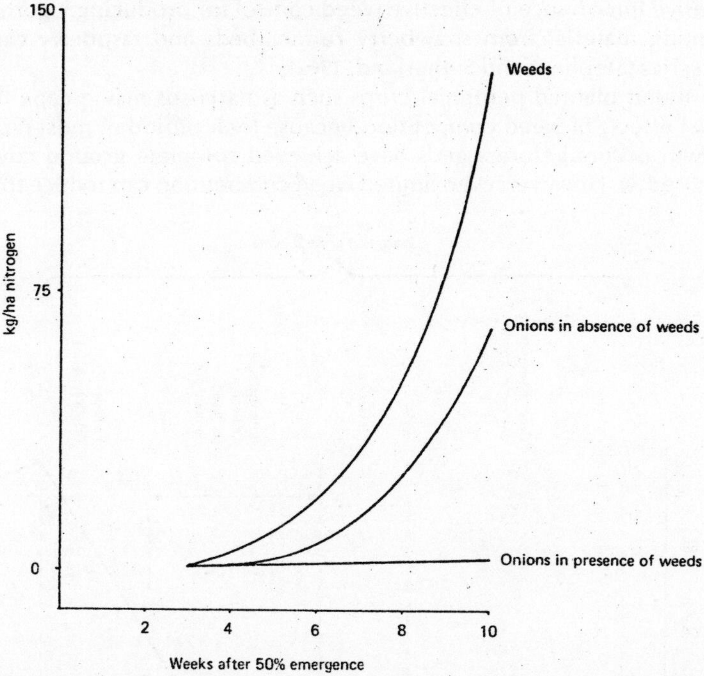


**Figure 1.3** Quantities of nitrogen removed from the soil by weeds alone, by weed-free onions and by onions in the presence of weeds (similar curves were obtained for potassium and phosphorus). (Adapted from Hewson and Roberts, 1973.)

spring they are immediately cut down to concentrate the plants' resources on cane production, and Lawson and Wiseman (1976) have shown that the critical period for weed competition in the newly planted raspberry occurs in early summer when the young canes emerge. Weeds left uncontrolled beyond this stage not only smother and kill some plants but reduce cane vigour, with serious long-term effects on the new plantation. In the strawberry the situation is less straightforward because weed competition in young plants during the early summer will not only reduce vigour and the numbers of new crowns produced but stolon and runner formation is also much reduced. In most years the effect of weed competition on vigour will be dominant, and the weaker smaller plants will be less productive. But in unusually dry years, the limited resources available may, in some cultivars, be devoted to stolons and runners instead of to fruit trusses; in these circumstances weediness may increase yields by preventing or reducing stolon formation (Chapter 10). These results emphasise the decisive importance of effective weed control for producing vigorous planting material from strawberry runner beds and raspberry cane nurseries (Stephens and Sutherland, 1962).

Autumn planted perennial crops such as narcissus may escape the worst effects of weed competition because their period of most rapid growth occurs before weeds have achieved complete ground cover (*Figure 1.4*). However, even limited weed competition can reduce total

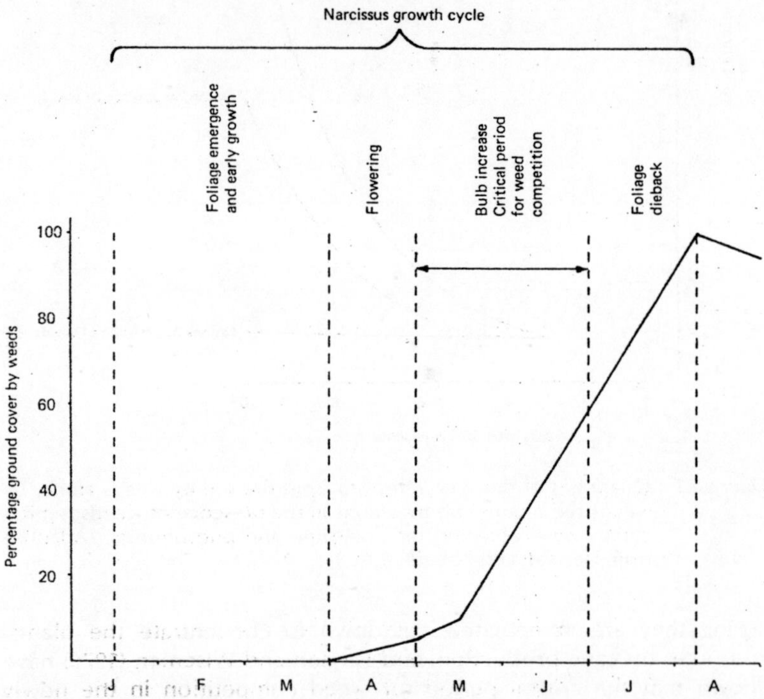


**Figure 1.4** The growth cycle stages of narcissus in relation to the increase in ground cover of annual weeds. (From Lawson, 1976.)

bulb weight and have appreciable effects on the size grading and on the potential for flower production (Lawson and Wiseman, 1978). In bulbs left in the ground for a second growing season it was found that the reduction in quality observed in one year crops was maintained even when weed competition in the second spring was prevented.

### 1.3.4 Competition and spacing

Maximum crop per unit area can only be achieved when the best use is made of water, minerals and incident light, and in practice this means there is competition between adjacent crop plants; any weeds present add to the total competition suffered by the crop. Weeds present in the rows of row crops are generally subject to early competition from the crop and are frequently smaller and less competitive than weeds between the rows (Bleasdale, 1973). It is common observation too, that crop suppression by weeds is greatest in weak and patchy crops and least in crops where germination and establishment are rapid. Ideally, weed control measures should be just one part of a management strategy designed to ensure fast and unhindered growth until the crop foliage provides a complete cover which suppresses further weed growth. Some crops when planted at a close spacing soon achieve good leaf cover, while others, including potato and red beet, take a longer time, offering weeds more opportunity to grow to a dominant position. A few crops, such as onion, can never provide a cover sufficiently dense to prevent weed growth and always remain poor competitors.

## 1.4 BENEFICIAL EFFECTS OF WEEDS

In addition to their well-known harmful effects, weeds can be beneficial. They enable areas of exposed soil to become rapidly clothed with a protective cover of vegetation, helping to prevent water or wind erosion, and on soils liable to wind erosion a cover of annual weeds is often encouraged for a time between the rows while the crop is developing its own cover. However, weeds may not always provide this protection, and when in parts of tropical Africa in which shifting cuitivation is practised the perennial Lalang-grass (*Imperata cylindrica*) become dominant it not only fails to give sufficient protection from erosion but also prevents the natural regeneration of woody plants that would provide an effective cover (Holm, 1976). Water weeds, which in excess cause increased water evaporation and restrict flow, are needed to maintain aeration and as a primary food source for fish. Weeds growing on unused land such as that adjacent to roads, fences and hedges provide food and shelter for parasites and predators of pests and for birds, some species of which eat large amounts of weed seeds in nearby fields; unfortunately, as such weeds may also act as sources of weed seeds, disease organisms and pests, they are liable to be destroyed indiscriminately. Weedy relatives of cultivated plants are sources of disease resistance and other useful variation for the plant breeder, and some may yet yield entirely new kinds of crop just as in prehistoric times non-shattering forms of wild oats growing as a weed among wheat grew more vigorously than wheat in the cold and wet far north so that farmers gradually found they were growing a new crop, oats.

## 1.5 WEEDS AS SOURCES OF PESTS AND PATHOGENS

### 1.5.1 Virus diseases and their vectors

Many specific relationships between weeds, crops and their pests and pathogens have been established, sometimes enabling action to be taken against weeds to prevent or reduce damage to crops. Insect and nematode pests which infect crops with viruses causing major diseases have often been shown to acquire the virus initially from weeds. For example, in the Salinas Valley in California, lettuce mosaic virus, which is seedborne in some batches of lettuce seed, can be spread rapidly within the crop by aphids, and in these circumstances weeds play only a minor role in the epidemiology of the disease. However, lettuce seed can now be tested for the presence of virus and only completely clean seed need be sown, but because the virus can be spread to the crop by aphids from nearby infected Sow-thistles (*Sonchus* spp.), weed control will in future become an integral part of the control programme for this disease. The relationship between crops and disease organisms is generally a complex annual cycle involving crops, vectors and weeds. In Arizona, for example, cucumber mosaic virus which causes epidemics of disease in cantaloupe melons, lettuce and sugar beet is carried by several weeds including species of *Brassica, Sisymbrium* and *Physalis*, by certain garden ornamentals and by several crops, with at least two aphid vectors. Not only do weeds act as a virus source but frequently also as the main reservoir of vectors. For example, many species of aphid alternate between an overwintering host from which winged migrants spread in spring and secondary hosts where colonies are established.

In parts of N.W. America, however, the peach aphid (*Myzus persicae*) overwinters not only on the peach in the egg stage but also continues in its summer viviparous form on weeds growing beside streams fed from warm springs, despite minimum air temperatures of –25°C. Experience shows that more aphids and a higher degree of infection with beet western yellows virus occurs in the fields nearest to these streams. This virus, which has now spread to Europe, has been found in N. America in weed populations well separated from beet crops, and its presence in them is thought to precede its spread to beet (Duffus, 1971); in Britain this disease is harboured by Groundsel (*Senecio vulgaris*) from which it is spread to lettuce by aphids. In England the milder winters frequently allow *M. persicae*, which acts as vector for many viruses, to overwinter in the viviparous stage on a variety of herbaceous hosts, where it provides the principal source of vectors for transferring virus from overwintering crops to spring sown brassicas (Broadbent, 1957).

In Britain raspberry ringspot virus produces a severe leaf curling and ultimately death in sensitive cultivars of raspberry such as Malling Jewel. In strawberry, this virus is usually combined with tomato black ring virus causing leaf mottling and severe growth depression and reduction in yield. Both viruses are soilborne, being transmitted by root sucking nematodes, but initially the means of carrying the disease over winter and after replanting from one crop to another was uncertain because the species of *Longidorus* nematode concerned could only retain infectivity for a few weeks. It was then found that Common Chickweed (*Stellaria media*), Groundsel (*Senecio vulgaris*) and several other ephemeral annual weeds carried the virus in a proportion of their seeds so that weed seedlings provided a continuing source for nematode reinfection (Cadman 1963; Murant and Lister, 1967). Nematode-transmitted viruses appear to be endemic in populations of wild plants, causing mild symptoms only; when a susceptible crop becomes infected the symptoms are generally much more severe. A similar pattern is shown by sow-thistle yellow vein virus which is widespread in plants of *Sonchus* spp., but is only locally important as a disease of lettuce.

Wild plants may not only act as original sources of viruses but having acquired new ones from nearby cultivated plants they may then act as local sources of the new virus. For example, the weed Nightshades, *Solanum nigrum* and *S. gracile*, have in Florida and elsewhere become infected with potato virus Y from potatoes, which can now be passed on to peppers, tomatoes, and of course, to succeeding crops of potatoes. Some viruses, such as celery yellow spot virus, cannot be spread directly by the vector within the crop, but instead must use an adjacent infected source, in this case Hemlock (*Conium maculatum*).

These examples illustrate the important part played by weeds in the population dynamics of vectors and in the epidemiology of virus diseases. In Scotland a combination of soil-acting herbicide and nematicide treatment has been successful in breaking the chain of reinfection with soilborne virus when a susceptible crop is replanted. Elsewhere weed control has frequently reduced the incidence of crop disease by removing sources of virus and vector.

### 1.5.2 Bacterial and fungal diseases

Weeds have also been implicated in the spread of bacterial and fungal pathogens, although their importance in this respect is less well defined. Many grass weeds, for example, can be found infected with the same kinds of fungi that cause severe diseases of cereal crops, but their significance as sources of infection is uncertain. Several cruciferous weeds can act as hosts of the organism causing club root disease in *Brassica* crops, *Plasmodiophora brassicae*, and it is also reported that zoospores of the same organism have been found in Yorkshire Fog (*Holcus lanatus*) although again their significance as disease sources is unknown. In the same way, grey mould fungus, *Botrytis cinerea*, occurs on flower heads of Sow-thistles (*Sonchus* spp.)

and this has been suggested as a possible source of infection for fruit crops, but without confirmation.

The best known cases of non-crop plants acting as sources of fungal pathogens are among the heteroecious cereal rusts, such as brown rust of wheat (*Puccinia graminis* var. *tritici*); together these rusts are thought to cause more than a 10 per cent reduction in world wheat yields. Brown rust alternates between two hosts, wheat and barberry (*Berberis vulgaris*), and it is on the latter host that the sexual phase occurs, in which segregation produces new races. A campaign designed to eradicate all *Berberis* and *Mahonia* bushes in the wheat areas of N. America is thought to have considerably reduced the ease with which new races of the rust are selected, thus making it easier for the plant breeder to match new rust races as they arise with new resistant cultivars. Although the disease has been reduced in severity, it cannot be eliminated because uredospores from diseased crops in Mexico are carried northwards in the air. This enables the organism to establish in the new crop without going through the alternate phase on barberry.

### 1.5.3 Nematodes

Weeds may act as alternative hosts of important nematode pests, an example being the potato root eelworm (*Globodera* spp.) which can persist in the soil in cyst form for many years, but can grow and reproduce on the roots of weed species of *Solanum* as well as those of the potato. Dormant cysts are stimulated to germinate by organic substances secreted into the soil by solanaceous plants, whereupon the young nematodes attach themselves to the host root and resume their growth cycle. A very similar situation occurs with the tobacco cyst eelworm which is also thought to reproduce on certain weeds.

### 1.5.4 Desirable effects of weeds on crop pests and diseases

Several examples have already been given of weeds providing food and shelter for aphids, which then attack crops, frequently also infecting them with viruses; however, the dynamic nature of pest populations, existing as they do in a shifting balance with parasites and predators also sheltering in vegetation, makes the true role of weeds hard to evaluate. Recent work suggests that some pests preferentially colonise pure crop stands, tending to avoid weedy ones. There is also considerable evidence that weeds act as hosts for beneficial predators and parasites, sometimes providing an overwinter food supply vital for their survival. It is argued that weeds should be encouraged provided they can be prevented from causing crop competition but managing weeds within such close limits is difficult (Altieri *et al.*, 1977; Altieri and Whitcomb, 1979; Van Emden, 1965).

REFERENCES

ALTIERI, M. A., SCHOONHOVEN, A. and DOLL, J. D. (1977). The ecological role of weeds in insect pest management systems, *PANS*, **23**, 185–206

ALTIERI, M. A. and WHITCOMB, W. H. (1979). The potential use of weeds in manipulation of beneficial insects, *Hort Science* **14**, 12–18.

ALTMAN, J. and CAMPBELL, C. L. (1977). Effect of herbicides on plant diseases, *Ann. Rev. Phytopathol.*, **15**, 361–385

BLEASDALE, J. K. A. (1963). Crop spacing and management under weed-free conditions, in *Crop Production in a Weed-free Environment*, Blackwell, Oxford, p. 114

BLEASDALE, J. K. A. (1973). *Plant Physiology in Relation to Horticulture*, Macmillan, London, p. 144

BROADBENT, L. (1957). *Investigations of Virus Diseases of Brassica Crops*, Cambridge University Press, Cambridge, p. 94

CADMAN, C. H. (1963). Biology of soil-borne viruses, *Ann. Rev. Phytopathol.*, **1**, 143–172

CHISAKA, H. (1977). Weed damage to crops; yield losses due to weed competition, in *Integrated Control of Weeds* (ed. J. D. Fryer and S. Matsunaka), University of Tokyo Press, Tokyo, p. 262

CRAMER, H. H. (1967). *Plant Protection and World Crop Production*, Bayer, Leverkusen, p. 524

DROST, D. C. and DOLL, J. D. (1980). The allelopathic effect of Yellow Nutsedge (*Cyperus rotundus*) on corn and soy beans, *Weed Sci.* **28**, 229–233

DUFFUS, J. E. (1971). Role of weeds in the incidence of virus diseases, *Ann. Rev. Phytopathol.*, **9**, 319–340

HARBORNE, J. B. (1977). Biochemical interactions between higher plants, in *Introduction to Ecological Biochemistry*, Academic Press, London, p. 243

HEWSON, R. T. and ROBERTS, H. A. (1971). The effect of weed removal at different times on the yield of bulb onions, *J. hort. Sci.*, **46**, 471–483

HEWSON, R. T. and ROBERTS, H. A. (1973). Some effects of weed competition on the growth of onions, *J. hort. Sci.*, **48**, 51–57

HOLM, L. (1976). The importance of weeds in world food production, *Proceedings 1976 British Crop Protection Conference – Weeds*, pp. 754–769

LAWSON, H. M. (1976). The effects of spring germinating weeds on narcissus, *Ann. Appl. Biol.*, **83**, 324–327

LAWSON, H. M. and SIEMAN, J. S. (1979). Competition between weeds and transplanted spring cabbage: effects of nitrogen top-dressing, *Hort. Res.*, **19**, 25–34

LAWSON, H. M. and WISEMAN, J. S. (1976). Weed control in spring planted raspberries, *Weed Res.*, **16**, 155–162

LAWSON, H. M. and WISEMAN, J. S. (1978). The effect of weeds on the growth and development of narcissus, *J. appl. Ecol.*, **15**, 257–272

MURANT, A. F. and LISTER, R. M. (1976). Seed transmission in the ecology of nematode-borne viruses, *Ann. appl. Biol.*, **59**, 63–76

NIETO, H. J., BRONDO, M. A. and GONZALEZ, J. T. (1968). Critical periods of the crop growth cycle for competition from weeds, *Pest Artic. C.*, **14**, 159–166

PARKER, C. and FRYER, J. D. (1974). Weed control problems causing major reductions in world food supplies, *FAO Plant Protection Bulletin*, **23**, 83–95

RICE, E. L. (1974). *Allelopathy*, Academic Press, New York, p. 353

ROBERTS, H. A., HEWSON, R. T. and RICKETTS, Margaret E. (1977). Weed competition in drilled summer lettuce, *Hort. Res.*, **17**, 39–45

SCOTT, R. K.. WILCOCKSON, S. J. and MOISEY, F. R. (1979). The effects of time of weed removal on growth and yield of sugar beet, *Agric. Sci. Camb.*, **93**, 693–709

STEPHENS, R. J. and SUTHERLAND, J. P. (1962). Effects of some herbicides on the weeds and crop in a raspberry cane nursery, *Hort. Res.*, **1**, 112–119

Van EMDEN, H. F. (1965). The role of uncultivated land in the biology of crop pests and beneficial insects, *Sci. Hort.*, **17**, 121–136

WELBANK, P. J. (1963). Toxin production during decay of *Agropyron repens* (Couch Grass) and other species, *Weed Res.*, **3**, 205–214

## FURTHER READING

ZIMDAHL, R. L. (1980). *Weed-Crop Competition – a Review*, International Plant Protection Center, Oregon State University, Corvallis, p. 196

# 2 THE ORIGINS, DISPERSAL AND CHARACTERISTICS OF WEEDS

## 2.1 THE ORIGINS OF WEEDS

The existing weed flora in any region is an amalgamation of indigenous species with introduced ones, although a clear distinction between the two cannot always be established because the origin of some weeds remains obscure. The provenance, for example, of Procumbent Yellow-sorrel (*Oxalis corniculata*), a weed of gardens and lawns in warmer regions, has yet to be definitely established, although it may have originated in the Pacific and spread to Europe by way of the Americas (Baker, 1972).

### 2.1.1 History of British Weed Flora

Our present weed flora in Britain is made up of a few hardy species that survived the ice ages, to which has been added a succession of invaders (Godwin, 1960). Among the weeds thought to have survived the harsh condition adjacent to the ice sheets are Groundsel (*Senecio vulgaris*), Common Chickweed (*Stellaria media*) and Small Nettle (*Urtica urens*), all species able to withstand sustained low temperatures. The first wave of aliens probably arrived in the late Stone Age, and it includes species able to survive and reproduce in cornfields, with seeds hard to separate from the grain, such as Charlock (*Sinapis arvensis*), Common Fumitory (*Fumaria officinalis*), Fat-hen (*Chenopodium album*) and Poppies (*Papaver* spp.). Other species now regarded as weeds were later introduced as food crops, for medicines, drugs or essential oils, as ornamental plants, or purely accidentally because they were caught up with animal fur, clothing or in packing materials. Ground-elder (*Aegopodium podagraria*), for example, was probably introduced because of its supposed therapeutic effect on gout, hence the alternative name, Gout-weed; but Ground-elder was also eaten as a pot-herb. In gardens it is a difficult weed to eradicate, and as Gerard recorded in a herbal of the 16th Century, Ground-elder or Herb Gerard is 'so fruitful in his increase that where it hath once taken root it will hardly be gotten out again ... to the annoying of better herbes'. The Romans were responsible for the introduction of many weeds of cultivation, whose dispersal must have been expedited by the extensive land clearance and disturbance while making roads and buildings. A weed unknown in Britain until this century is the Slender Speedwell (*Veronica filiformis*), a native of Asia Minor, which was first introduced as an attractive blue-flowered rock garden plant. Since about 1925 it has escaped from gardens and has become a common weed of turf that is difficult to control because it is resistant to most selective herbicides. Dispersal has been almost entirely vegetative by the scattering and rooting of small stem fragments, but the plant can produce seed although it probably rarely does so in Britain (Salisbury, 1961).

### 2.1.2 Prevention of new weed introductions

In recent times increased trade and travel between countries and continents has accelerated the rate at which plant species have spread through the world. The same vigilance is required to prevent the further spread to new areas as is exercised, for example, to prevent the establishment in Britain of the Colorado beetle or the rabies virus. But because a plant may be harmless in one environment yet become a serious problem weed when taken to a new one, or on the contrary, a problem weed in one country may fail to show its full potential when taken to another, workable international regulations to prevent further spread of weeds will be hard to formulate.

### 2.1.3 Weeds as opportunists

Our worst weeds are those associated with cultivation and man has probably been bothered by them since he first cultivated land. As soon as the existing natural vegetation is cleared leaving the soil bare, a new set of plants, taking advantage of the space and the lack of competition, can swiftly build up a large population covering all the available bare areas from a few pioneer plants. Such vigorous, adaptive plants are familiar to farmers and growers everywhere as the weeds of arable cropping. A similar group of species are known to the botanist which by their opportunism quickly cover naturally occurring open situations such as sand dunes or those produced by volcanic eruption or landslides. The arable weeds are sometimes called agrestals and those on natural sites ruderals although, in practice, a similar group of species are involved. Some agrestals have responded to selection pressure by evolving specialised forms which mimic the crop either in seed anatomy or growth form and may adapt closely to its life cycle. Areas cleared for cultivation may also inadvertently provide conditions that favour the establishment of an alien weed; in Australia, for example, Skeleton-weed (*Chondrilla juncea*) probably cannot establish in competition with the native vegetation, gaining a foothold only on cleared or cultivated areas.

There was originally little difference between crops and weeds because the first crops were probably mixtures of plants able to produce harvestable seeds. Remarkably well-preserved bodies of Iron Age men and women have recently been recovered from the acid peats of Northern Europe. In the gut of one such body, known as Tollund Man, were found remains of a variety of seeds including barley and flax with Pale Persicaria (*Polygonum lapathifolium*) and smaller amounts of Docks (*Rumex* spp.), Black-bindweed (*Polygonum convolvulus*), Gold-of-pleasure (*Camelina sativa*) and Green Bristle-grass (*Setaria viridis*) (Glob 1969; Helbaek, 1953). This and other evidence suggests that ordinary people at this time consumed a gruel made from such a mixture of seeds from crops, as we now understand the term, and of what are now common arable weeds.

### 2.1.4 Aggressive properties of weeds

A property possessed by all successful, or from man's point of view harmful and noxious, weeds is that of aggressiveness. This defies definition and there is no satisfactory way of measuring it but no farmer or gardener would deny its existence! The number of really harmful or aggressive weeds is quite small, being only about two hundred according to Holm (1976), representing a minute proportion of the total named world flora. So what qualities do these plants have in common to make them so special? It is frequently found that the worst weeds in an area are aliens. Perhaps it is inevitable that in Britain, where the entire flora became severely depleted in successive ice ages, that most of the weeds should be aliens, filling the many types of 'niche' left vacant. But what of countries with a rich native flora, such as Australia, New Zealand or the Americas? Here again it is generally found that the greatest weed problems are posed by introduced species, which in many cases are either not weeds at all or are of only local importance as weeds in their country of origin. For example, Willows (*Salix* spp.) in New Zealand grow so rampantly in wet areas that they dam up streams causing flooding. The Prickly Pear (*Opuntia inermis*), introduced into Australia as an alternative source of cattle fodder, grew completely out of control and covered almost 30 million hectares of good land to a height of 2 m before it succumbed to biological control (Chapter 5). The Water Hyacinth (*Eichhornia crassipes*) is a native of the Amazon basin where it causes no particular problem, being in balance with other plant and animal life, but carried to Central and North America, Africa and Asia it has caused havoc in the running water of canals, ditches and rivers. However, not all non-weedy plants require to be taken to a new site for them to achieve weed status; weeds can arise *in situ*. In Britain the Rosebay Willowherb (*Epilobium angustifolium*) was until this century a fairly uncommon plant, occasionally planted in gardens but generally only seen in woodland clearings and on heaths following fires where it apparently thrives on the high mineral status of the soil which may be shunned by other species; hence its alternative name, Fire-weed. Perhaps because of an increase in tree felling and of accidental woodland fires, coupled with the plant's prodigious seed production, the population of Rosebay Willowherb has increased explosively this century. Each plant can produce 100,000 or more seeds equipped with small parachutes which ensure effective long-range seed dispersal, and the seedlings grow vigorously, spreading by means of shallow rhizomes. Yet for reasons quite unknown, this weed of the twentieth century formerly showed none of its potential for aggressive spread.

The eventual outcome when a new species is introduced in sufficient numbers for possible establishment will depend upon the interplay of many factors. With just two species grown together experimentally, there may sometimes be mutual stimulation or depression of growth, and even if one of the two partners grows less vigorously than the other when cultured alone, it may yet oust the potentially more vigorous partner when the two are grown together. Even more when a species is introduced to an entirely new environment must its eventual success or failure depend on a complex interaction of biotic and abiotic factors, and it should come as no surprise that the great majority of introduced aliens maintain no more than a tenuous foothold in their new home; a much smaller number find conditions favourable, perhaps because of the total absence of some biological control agent or for other reasons, and the population increases explosively. After a time, perhaps many years later, the indigenous plants and animals come to terms with the invaders, exploiting its weaknesses perhaps, and the original high populations fall. Canadian Waterweed (*Elodea canadensis*) was first recorded in England in 1847 near Market Harborough, then at the centre of the British canal system. Spread was rapid and 'it multiplied to a state of excessive profusion' in ponds, lakes, canals and rivers. Since that time its dominance has declined equally dramatically until today it is common but seldom abundant. A recent parallel may be occurring in Africa, where the originally rapid growth of Kariba Weed (*Salvinia molesta*) on Lake Kariba now shows signs of declining (Gaudet, 1979).

## 2.2 DISTRIBUTION AND SPREAD OF WEEDS

### 2.2.1 Problems faced by an invading species

With the help of man many alien species can be maintained in gardens but it is fortunate that few are able to survive unaided in the wild when in competition with other plants. Hundreds of aliens have been recorded near docks and other places where foreign goods, especially wool, are handled, yet most soon disappear unless reinforced with fresh seed from abroad. The chances against success for a newcomer are considerable but those that do 'hit the jackpot' (Baker, 1974) are often highly successful. This may be directly, or indirectly by forming a vigorous hybrid with a native species. This occurred in the nineteenth century when the Smooth Cord-grass (*Spartina alterniflora*), introduced from North America, formed a sterile but strong growing hybrid with the native grass *S. maritima*; following chromosome doubling the new and fertile species *S. anglica* (now called Common Cord-grass) was created, which is so vigorous that it has largely ousted the native grass. Although sometimes planted to bind estuary mud, in other situations the hybrid grass is an aggressive weed.

Newly introduced weeds are often quite unrepresentative of the whole species, because they generally come from restricted, local populations (Harper, 1965) and the aggressive form of an alien weed therefore may be biologically untypical of the original form, differing perhaps in herbicide susceptibility, and in reaction to different strains of any proposed biological control agents. The Slender Speedwell (*Veronica filiformis*), already referred to, probably sets no seed in Britain because the plant is self-sterile, and our infestation appears to be derived from a single strain of the weed, perhaps from a single original introduction.

### 2.2.2 Man's part in weed introductions

Introduced weeds often enter a country accidentally and only after the plant has become well established and its danger has been recognised are attempts made to eradicate it. *Solanum elaeagnifolium*, a plant from the Southern United States, was recently introduced to South Africa, probably in the form of seeds present in imported pig feed. This plant, which is now known in South Africa as Satansbos (Devils-wood), is spread by seed carried in water or blown by wind. The plants are said to supplant all other vegetation, and are merely spread further by cultivation. In addition it is poisonous to livestock, and now that its danger is well understood control is being attempted and the public have been asked to cooperate by reporting all new outbreaks. Another plant from North America, *Parthenium hysterophorus*, was accidentally introduced to Queensland, possibly in contaminated grass seed in about 1958, but as its initial spread was slow little immediate alarm was shown. However, since 1973, after unusually heavy rains and mild winters it has spread rapidly and alarmingly (Haseler, 1976). Initial distribution of seeds was by traffic along roads, the seedlings thriving beside roads where water and mud accumulate, but it is extending into pasture land, supplanting the native grasses, and now occurs in patches of up to 10,000 plants per hectare; it is feared that it may now spread to areas of arable farming. This plant is notorious in India (Khosla and Sobti, 1979), where it has also been recently introduced, as a cause of severe contact dermatitis in man and horses, but in North America it is a lesser problem because it competes poorly there and is generally found only in open habitats. Each plant produces about 15,000 seeds giving an enormous potential for increase and active steps, including a search for biological control agents, are being taken to contain the pest.

All too often introductions have been carefully planned to give the new plant the best chance of succeeding. Early settlers from Europe established naturalisation societies for this purpose, and some of their results can be seen with the pest status of the rabbit in Australia, and of Gorse (*Ulex europaeus*), Willows (*Salix* spp.) and Bramble (*Rubus fruticosus*) which have all become weed problems in New Zealand. After a few years habituation in botanic gardens, many plants have either been deliberately introduced to the wild or been allowed to escape. Gallant Soldier (*Galinsoga parviflora*) has become a weed of cultivation in market gardens in Southern Britain, having escaped from Kew Gardens, London, in the nineteenth century. The common name, a corruption of the Latin generic name, arose because there was no existing common British name for this alien from South America. Botanic gardens have also been blamed for the escape of several harmful aquatic weeds such as the Water Hyacinth (*Eichhornia crassipes*) in Africa, first to the Congo then to the Nile.

There have also been some notable escapes from gardens, such as Slender Speedwell in Britain. A more serious problem has been caused by the introduction around the World of several attractive species of *Oxalis*. The pink-flowered *O. latifolia* from Central America increases by means of numerous small bubils, which easily become detached and are distributed in the soil adhering to plant roots. It has become a considerable nuisance in nurseries, glasshouses and private gardens in Jersey and Southwest England, as well as being a weed in the Mediterranean area, in South Africa and in Sri Lanka. Another similar species, the Bermuda Buttercup (*O. pes-caprae*) is present in the bulb fields of the Scilly Isles but is insufficiently hardy for mainland Britain. However, the spread of this species to North Africa and Australia has created almost insoluble problems there because the bulbil-forming *Oxalis* species are impossible to control by cultivation, which only spreads them further, and they are resistant to most herbicides. The weed not only monopolises the land, but the foliage is poisonous to cattle. Had greater care been taken to prevent these garden plants escaping and becoming naturalised an immense expense could have been avoided, but gardeners could not be expected to have foreseen this. Some plants introduced as new crops then become aggressive weeds. *Opuntia inermis*, the Prickly Pear, introduced as a crop to Australia has been mentioned. Now regarded as a major world weed, Johnson-grass (*Sorghum halepense*) was first introduced to the Southern United States from Europe as a forage grass, but its vigour, rapid seed production and rhizome production have led to its establishment as a competitive weed throughout the tropics. Care has to be taken when growing seed of herbage sorghums to prevent cross pollination with Johnson-grass, which produces weed types rather in the same way that annual beet can appear in sugar beet seed.

The greatest single source of weed introductions is through impure crop seed. In advanced countries commercial seed lots are routinely sampled and tested for weed seed contamination; but in the past this was not so and foul seed was the cause of weeds being introduced to the very place where they could do most harm, namely with the crop to whose growth cycle they had long become adapted. It is virtually impossible to ensure freedom from weed seeds and Horne (1953) has shown that levels of contamination that are barely perceptible to the eye can lead to large numbers of weeds in the field. For example, Cleavers (*Galium aparine*), present as 0.1 per cent of the weight of wheat sown at 170 kg/ha would, if uniformly distributed, give between 3 and 4 seeds/$m^2$. Where it is imperative to prevent weeds spreading to an uninfected area, every viable weed seed is important, and the EEC regulations on seed quality allow a maximum of 1 Wild-oat (*Avena fatua* or *A. ludoviciana*) per kg of cereal seed. Farmers may often be their own worst enemies when they sow untested seed bought from other farmers; recent surveys in Britain have shown an alarming incidence of wild oats in farm seed drills.

### 2.2.3 Natural dispersal of weed seeds

Many plants are adapted for wind or water distribution of their seeds and several have evolved mechanisms for dispersing their seeds explosively. Fruits may be attractive to animals, which then distribute the seeds; some seeds actually pass unharmed through the animal's gut (Piggin, 1978). Seeds and fruits may also be modified for clinging to or even penetrating the fur of animals, and these seeds cling also to clothing and hessian sacks. There have been suggestions that migrating birds carry seeds between continents, and coconuts can certainly drift from considerable distances; new volcanic islands and land reclaimed from the sea soon become clothed with plants. But the record of man's involvement in the introduction of new species suggests that most natural mechanisms for plant dispersal are essentially local and short range.

The annual 'snowstorm' of Rosebay Willowherb seeds, followed by colonisation of all suitably open sites, is testimony to the effectiveness of wind dispersal. The parachutes remain open, continuing to drift with the wind, in dry air; moisture makes the parachute close and drop to the ground. Similar parachutes are found in many weed members of

Compositae, such as Dandelion (*Taraxacum officinale*), the Sow-thistles (*Sonchus* spp.), Coltsfoot (*Tussilago farfara*), the many weed species of *Senecio* and the various thistles (Burrows, 1974). In Creeping Thistle (*Cirsium arvense*) the parachutes are usually barren and if seeds are set they adhere together and fail to disperse. Occasionally, however, seeds of this species do become airborne, and together with Colt's-foot it has been one of the pioneer species on reclaimed polders in Holland. Some of the windborne tree seeds, including Birches (*Betula* spp.) and Willows (*Salix* spp.) can be a nuisance in container nurseries where they find conditions ideal for establishment. Another form of wind dispersal is shown by the North American Tumbleweeds (*Amaranthus graecizans* and *A. albus*). At maturity the plant becomes spherical, breaks off at ground level and rolls over the plains of N. America, scattering seed as it goes.

Another effective mechanism for short range dispersal is that of explosive fruits, sometimes able to shoot seed for several metres. The Hairy Bitter-cress (*Cardamine hirsuta*) has become a universal weed of intensive horticulture, showing exceptional plasticity in its ability to produce a few seeds from minute, starved plants, but also to produce up to 50,000 seeds from large well-nourished specimens; each can be shot as far as 60 cm away from the parent plant (Salisbury, 1961). Gorse (*Ulex europaeus*), the annual weed Spurges (*Euphorbia* spp.), some *Oxalis* spp. and the Indian Balsam or Policeman's Helmet (*Impatiens glandulifera*) also have effective explosive seed dispersal. The Dwarf Mistletoes (*Arceuthobium* spp.) are minute tree parasites, which despite their small size do considerable damage to timber and other tree crops (Chapter 4). The seeds are shot from exploding fruits, scattering most seeds within 1 or 2 metres but exceptionally to almost 14 m (Hudler and French, 1976).

Weed seeds may also be adapted for transport in water, and in the United States an astonishing total of 150 different kinds of seed were recovered from one irrigation main. The numbers of seeds were not large compared with the potential number of new seeds produced in a weedy field, but 100 seeds/m$^2$/annum is still quite large enough to vitiate the effect of other control measures, and the screening of irrigation mains is recommended (Kelley and Bruns, 1975). But perhaps the real significance of waterborne weed seeds is qualitative rather than quantitative, because irrigation water is the perfect medium for distributing the seeds of a weed new to the area.

### 2.2.4 Weed Evolution

Evolution of living organisms is a continuing process although it is sometimes wrongly considered to have begun and ended in the past. Since arable agriculture began, weeds have evolved to form a close relationship with specific crops, by copying their growth form or life cycle or by producing seeds or fruits that cannot easily be separated, and this long period of adjustment may be one reason why so many old world weeds have been successful when introduced into the Americas, Australasia and Africa. The closer the genetic relationship between crop and weed the harder it may be to control the weed. Red rice and wild oats, weeds of rice and temperate cereals, respectively, are examples of such close affinity causing severe problems of weed control. Differences between wild oats and either wheat or barley have, however, proved sufficient for selective herbicides to be used effectively but these are expensive and wild *Avena* species remain a serious threat to cereal production (Price Jones, 1976). The red rice problem is caused by wild and unproductive forms of cultivated rice (*Oryza sativa*) or by closely related *Oryza* species, all of which have small, hard, pigmented fruits in inflorescences which generally shatter early. One ingenious solution to the problem in India has been to breed a purple anthocyanin marker gene into the crop which enables the red rice with normal green leaves to be easily identified and removed by hand, leaving behind the crop with its streaks of red in the foliage. This type of solution, relying on differentiation between crop and weed by human eye is, however, not appropriate for extensive cropping.

### 2.2.5 Evolution by hybridisation

In the evolution of plants, hybridisation between related species, chromosome doubling and other genetic changes have produced new sorts of plants including some weeds (Stebbins, 1971). Common Hemp-nettle (*Galeopsis tetrahit*), for example, is believed, from chromosome studies, to have arisen from hybridisation between two *Galeopsis* species, *G. pubescens* and *G. speciosa*, followed by chromosome doubling of the hybrid.

Plants that suddenly become weeds have often proved to be of hybrid origin. In California, for example, an aggressive arable weed that has spread northwards along the Pacific coastline has arisen by hybridisation between the Cultivated radish (*Raphanus sativus*) and Wild Radish (*R. raphanistrum*).

The Water Fern, Kariba Weed or African Pyle, which has recently caused so much trouble on lakes, rice paddy and in irrigation canals in Africa and Asia, was first considered to be *Salvinia auriculata*, a relatively harmless plant of the tropical lowlands of S. America. It now appears likely that this weed is a new and extremely vigorous species derived from hybridisation between two other South American *Salvinias* species, for which the apt name of *S. molesta* has been coined (Mitchell, 1972).

Probably the greatest potential for new problems is the evolution of weed forms of crop plants. These may arise by hybridisation between crops and closely related weeds during seed production, then become established in the field, where they are especially difficult to control because there may be no way of differentiating between crop and weed at the seedling stage. A problem of this type has arisen in the sugar beet growing areas of Northern Europe: annual beet is a weed type that bolts in the first year without producing a swollen root (Longden, 1976; Hornsey and Arnold, 1979). Several cultivars of sugar beet are monogerm triploids produced by pollinating male sterile monogerm diploid mother plants with pollen from adjacent rows of tetraploids. It has been shown that wild beet diploids release their pollen each morning some two hours earlier than the tetraploid so there is a period when any pollen in the air will be from wild diploid plants. In Southern Europe, where some of the beet seed is grown, a high proportion of the local diploid beet are annuals, so that any hybrid seed produced by pollen contamination will contain annual bolters. The worst kind are the early annual beets that flower from June onwards since these not only produce about 150 seeds per plant, ensuring their repeated future presence in that field, but their pollen, carrying the factors for the annual habit, is available for crossing with and thus the contamination of nearby seed crops. The affected acreage in Britain is small, but locally important, while some fields in Europe have been taken out of beet production. Regulation of the problem is achieved by testing all seed lots before release and rejecting any that contain annual beet, and also by attempting to prevent any increase in fields already infested. Successful control in the field has been achieved by hand pulling, mechanical removal of beet inflorescences and by application of glyphosate to the weed beet plants (Longden, 1976).

The potential for a parallel situation exists with the small fruited melon or Dudaim (*Cucumis melo* var. *dudaim*) which is already a serious weed of asparagus in the Imperial Valley of California, and could become a threat to edible melon (*Cucumis melo* var. *reticulata*) production, both as a weed and as a source of contaminating pollen in melon seed production. Strenuous efforts are being made to prevent the dissemination of the small weed melon fruits or pepos (Chapter 4).

## 2.3 SPECIAL QUALITIES OF SUCCESSFUL WEEDS

### 2.3.1 The weed families of plants

When weeds are defined as plants in the wrong place, it follows that any kind of plant, including crops, can at some time qualify for weed status. However, few kinds of plant consistently survive man's attempts to suppress them and of the many thousands of named plant species only 206, according to Holm (1976), can be regarded as weeds important in human affairs and of these a mere 80 are of major world significance. Although Holm lists 59 plant families that contribute important weeds, two thirds of his 206 kinds come from only 12 families (*Table 2.1*). Three of these, Gramineae, Compositae and Cyperaceae, between them contribute almost half of all the world's major weeds (43 per cent). When the list is divided into what Holm calls primary weeds and secondary ones, then these same three families are seen to include almost 60 per cent of all primary weeds. Baker (1974) emphasises that certain families, including Amaranthaceae, Cyperaceae and Malvaceae are of more importance in the tropics, whereas Compositae, Cruciferae, Polygonaceae and Umbelliferae are mainly confined to temperate regions. With the notable exception of Compositae, the 'weed families' also provide most of the world's food crops, with Gramineae (the grasses) providing most of the crops as well as many weeds. In addition to the families listed there are almost 50 more otherwise innocuous ones each containing a few noxious weed species.

*Table 2.1*

*Flowering plant families contributing important weed species. Forty-seven other families contribute either 1, 2 or 3 species. (From Holm, 1976.)*

| *Family* | *Number of species* | | |
|---|---|---|---|
| Gramineae | 44 | 43% | 68% |
| Compositae | 32 | | |
| Cyperaceae | 12 | | |
| Polygonaceae | 8 | | |
| Amaranthaceae | 7 | | |
| Cruciferae | 7 | | |
| Leguminosae | 6 | | |
| Convolvulaceae | 5 | | |
| Euphorbiaceae | 5 | | |
| Chenopodiaceae | 4 | | |
| Malvaceae | 4 | | |
| Solanaceae | 4 | | |

### 2.3.2 Properties of weeds

What special properties are shared by these remarkable plants? Several authors (Baker, 1974; Hill, 1977) have listed features that appeared to them to have been crucial to the success of weeds. These can be summarised under the next five headings.

*High degree of plasticity*

Annual weeds generally have the ability to mature and produce at least a few seeds under conditions too hostile for completion of the life cycles of most other plants. When conditions are more propitious weeds flourish and produce larger numbers of seeds. Uprooted annual weed seedlings, such as Groundsel (*Senecio vulgaris*), can often complete their development and produce viable seeds although deprived of water. Weed seedlings germinating in late summer commonly compress their life cycle into a few weeks, producing viable seeds from minute plants 1–5 cm tall. Such plasticity of form which enables weeds to mature and produce seed in a wide range of environments and which has already been noted for Hairy Bitter-cress (*Cardamine hirsuta*), can be contrasted with the more exacting requirements for growth and reproduction of crops and other non-weed species. Baker (1965) refers to 'general purpose genotypes' possessed by successful weeds, providing an overall fitness for differing environments, but without the finely tuned adaptation to soil and climate shown by the locally adapted species that soon supplant weeds when cultivated land is abandoned. Weeds, however, may have precise temperature or day-length requirements for germination and flowering, respectively, especially those weeds whose development is synchronised with that of particular crop plants. Winter annuals, such as Cleavers (*Galium aparine*), the Winter Wild-oat (*Avena ludoviciana*) and Ivy-leaved Speedwell (*Veronica hederifolia*) germinate mainly in the late autumn and winter in Northern Europe, when it is relatively cold, eschewing the warmth of spring favoured by most other weeds. This puts winter annuals into a position where they can compete with autumn sown crops. Such behaviour betrays their Mediterranean origin, where growth occurs in the cool and wet of the winter rather than in the summer when it is prevented by drought. Other species, such as the Wild-oat (*Avena fatua*) and Knotgrass (*Polygonum aviculare*), require the cold of winter to overcome dormancy, so that their seeds germinate in spring, and they compete with spring sown crops (Chapter 3).

A further aspect of plasticity is tolerance of levels of nutrients in the soil that would be too high for other plants, and the apparent predisposition of Rosebay Willowherb (*Epilobium angustifolium*) for freshly burned areas with high potassium has been referred to. Several arable weeds, including Fat-hen (*Chenopodium album*), have long been associated with rich media such as soil near manure heaps, and probably in ancient times with the nutrient-rich garbage heaps adjacent to man's dwellings.

### *Capacity for rapid increase*

When soil is cultivated after being in pasture for many years, any weed seedlings that develop are either from the reserve of dormant seeds in the soil or from seeds that have been recently brought or blown in. In either case the number of pioneer weeds may be small. Most annual weeds are self-fertile, which enables a single plant to build up a large population rapidly. Ephemeral annuals, that is weeds having a short life cycle and producing several generations each year, have the potential for rapid increase. Common examples in Britain are Annual Meadow-grass (*Poa annua*), Groundsel (*Senecio vulgaris*) and Shepherd's-purse (*Capsella bursa-pastoris*). Some species ensure a rapid increase from the pioneer plants by dispensing with the sexual phase in seed production; seeds produced via this apomictic route give rise to seedlings that are identical to the parent plant. Occasionally normal sexual reproduction introduces variation so that apomictic species such as Dandelion (*Taraxacum officinale*) exist as numerous clones which may be visibly different from each other. This is a similar situation to that found in Common Couch grass (*Agropyron repens*) where the population exists as a series of clones, but in this case asexual reproduction is by growth and dispersal of rhizome fragments rather than by seed. Not only are there visible differences in habit or form between such clones but they may differ also in herbicide susceptibility. Repeated selfing in self-pollinating species induces a high degree of homozygosity, of uniformity among the offspring and to the establishment of distinct local strains. The end result may be visibly similar to that caused by seed apomixis or other types of vegetative reproduction; in either case clones or races with differing physiological or edaphic requirements ensure that each weed species can colonise a range of habitats.

In many troublesome perennial weeds, success depends not only on vegetative spread but also on prolific seed production. Familiar examples include Rosebay Willowherb (*Epilobium angustifolium*), both common British Dock species (*Rumex obtusifolius* and *R. crispus*) and the tropical Johnson-grass (*Sorghum halepense*). There are other weeds, however, in which seed production plays but a minor role and increase is largely by vegetative means.

*Ability to produce seeds with discontinuous germination patterns.*
The cultivated Shirley Poppy is derived from the same species (*Papaver rhoeas*) as the common red poppy of cereal fields, yet the two show very different germination patterns. Most seeds of the cultivated poppy can be expected to germinate within a few days of sowing, but only a small proportion of wild poppy seeds will germinate at any one time; most of the remainder being viable but dormant. Such dormancy is a normal feature of most wild plants but in cultivated forms dormancy has been selected against. The effects of various types of dormancy, both innate and arising from environmental factors, ensure that a small part only of the total viable seed population is capable of germination even when conditions of temperature, soil aeration and water supply are favourable. Some species, including Darnel (*Lolium temulentum*) and Corncockle (*Agrostemma githago*) have achieved importance as weeds without marked discontinuities in their pattern of germination. Both species show little innate or environmentally induced dormancy, most seeds germinating readily when conditions

permit, and they can therefore only be maintained as weeds of cultivation by constant replenishment from contaminated crop seed. In this way Darnel and Corncockle survived in Britain as frequent weeds of cereals but more effective seed cleaning has eliminated them during the last one hundred years.

Aggressive weeds also owe much of their success to effective means of dispersing seeds or vegetative propagules. Long-term control measures depend for their effectiveness on an understanding of the dynamics of increase and dispersal because only then can control be concentrated on the most vulnerable phases (Chapter 3).

*Ability to compete with other plants*

Unless treated with herbicides, mown grass swards contain, in addition to grasses, a varied flora of broad-leaved plants. Some are rosette forming species, including Daisy (*Bellis perennis*), Dandelion (*Taraxacum officinale*) and various Plantains (*Plantago* spp.), the leaves of which grow just above the grass and suppress it, while remaining below mowing height. Others such as Common Mouse-ear (*Cerastium holosteoides*), Selfheal (*Prunella vulgaris*) and Creeping Buttercup (*Ranunculus repens*) produce vigorous leafy horizontal stems or runners, which locally suppress the grasses and establish weed patches. These examples of effective competition with desirable species illustrate the need for weeds to possess a competitive edge. The means of competition is obvious in rosette plants in turf, with scrambling climbers such as Field Bindweed (*Convolvulus arvensis*) and with parasitic weeds, but in general the suppression evident in weedy crops results from the combined effects of several factors acting to the weed's advantage.

*Effective means of vegetative reproduction*

Perennial weeds, whether or not they produce and disperse their seeds are at an advantage if they can also reproduce vegetatively. This can be by runners, stolons or rhizomes, or bulbs, corms or bulbils. Many weeds regenerate from small pieces of stem, root or rhizome cut or broken during mowing or cultivation. Occasionally, as for example with Slender Speedwell (*Veronica filiformis*) in Britain, spread has been largely vegetative probably because this self-sterile species seldom sets seed. Seeds of Creeping Thistle (*Cirsium arvense*) can germinate and begin a fresh infestation but this dioecious species appears not to produce seed prolifically and spread is mainly by growth and dispersal of buds produced on the brittle swollen roots. In Common Couch (*Agropyron repens*) the relative importance of reproduction by seed and by rhizomes is less obvious. Couch Grass is self-sterile, pollen from another clone being essential for fertilisation to occur. Seed set is therefore variable, being high only when two or more clones grow fairly close together. Probably the greatest importance of seed in this species is as a crop seed contaminant, which can introduce Common Couch to a previously clean field; spread within a field is apparently mostly vegetative, perhaps because the seeds lack both dormancy and a dispersal mechanism, and probably remain viable for no more than one or two years. Spread of the bulbil-forming *Oxalis* species is mainly by breaking up and dispersing the groups of bulbils by cultivation; rotary cultivation soon provides a dense and uniform carpet that competes strongly with crop seedlings.

Some weeds, such as the Docks (*Rumex* spp.) and Dandelion (*Taraxacum officinale*), grow readily from pieces of underground stem broken during cultivation. Both Docks and Dandelions, however, also produce seed and spread is mainly by seed dispersal. The vigorous Horse-radish (*Armoracia rusticana*) does not set seed in Britain and therefore spread in gardens and along roadsides is solely by extension growth and dispersal in top soil of its swollen storage roots.

Another species spread in this way is Japanese Knotweed (*Polygonum cuspidatum*) which each year forms a thicket of cane-like stems growing up to two metres tall. Both the Japanese Knotweed and Horse-radish often occur beside roads as gradually extending isolated patches.

**Photosynthesis by C3 and C4 pathways in crops and weeds**

Some plants have been found to possess a modified leaf anatomy, characterised by the photosynthetic tissues being arranged in concentric rings of cells containing chloroplasts around the leaf veins. The significance of this remained obscure until plants with this cell arrangement were shown to be adapted physiologically to make the best use of bright sunlight and high temperatures, while experiencing a limited water supply. Such plants are now known as C4 plants because the early products of photosynthesis, malate or aspartate, contain four carbon atoms. In contrast, plants adapted to growth in cooler regions, and all plants which do not have the leaf anatomy of C4 species, are known as C3 plants because the first compound formed is phosphoglyceric acid which contains three carbon atoms. C4 anatomy and physiology have been found in maize, sorghum, sugar cane and millet, in many weed grasses such as Johnson-grass (*Sorghum halepense*), Bermuda-grass (*Cynodon dactylon*) and Cockspur or Barnyard-grass (*Echinochloa crus-galli*), and also in some weed members of the genera *Amaranthus, Atriplex, Portulaca, Salsola* and *Cyperus*. The possession of C4 physiology appears to confer no advantage in cool climates although it is interesting that the aggressive and successful hybrid, Common Cord-grass (*Spartina anglica*), which occurs on salt marshes is a C4 plant. Areas providing conditions advantageous to C4 land plants occur well beyond the tropics; for example, in summer the entire United States mainland is favourable for them and C3 crops such as rice, soybean, tobacco, sunflower and cotton may well be at a disadvantage when confronted with C4 weeds (Black *et al.*, 1969).

## 2.4 THE CHANGING WEED FLORA

Weeds present at any one time do not reveal the total potential for weediness because soils contain many dormant but viable weed seeds or vegetative buds. Production of weed seeds can have large and long-term effects on future weed populations. For these reasons trends in the composition of the weed flora can only be followed quantitatively, except in rare instances where species become locally extinct.

Changes in the relative abundance of different species or of sub-specific variants can be caused by the interplay of management and climatic factors including crop rotation, sowing date, timing and nature of cultivations, by liming and manuring and, more recently, by the action of herbicides. Species with dormant seeds are buffered against permanent change, but those without this protection can become reduced or eliminated quite rapidly. The earlier mentioned rapid decline of Darnel (*Lolium temulentum*) following improved seed cleaning appears to have been accelerated by a lack of seed dormancy in Darnel. The introduction of herbicides has been accompanied by other changes so that effects due to herbicides alone cannot easily be isolated but studies have been made of trends in the composition of weed floras since herbicides came into general use (Chapter 8). Most reports have shown a decline in the proportion of susceptible species and an increase in resistant ones, especially of grasses where MCPA and 2,4-D are regularly used. In sugar cane fields repeated application of the phenoxy-acetic acid herbicides caused an increase in grass weeds, but now the dominant weeds are often Sedges (*Cyperus* spp.) following regular use of grass killing herbicides. Fryer and Chancellor (1970) reported similar trends in Britain, with Common Couch (*Agropyron repens*), Annual Meadow-grass (*Poa annua*) and Wild-oat (*Avena fatua*) increasing and the formerly dominant cereal weeds Charlock (*Sinapis arvensis*) and Common Poppy (*Papaver rhoeas*) much reduced. However, some broad-leaved species are now more prevalent, especially Common Chickweed (*Stellaria media*), Knotgrass (*Polygonum aviculare*) and other *Polygonum* species, some Speedwells (*Veronica* spp.) and Fat-hen (*Chenopodium album*).

During the past 25 years there has been a transition in British horticulture from reliance on intensive methods with much hand work to a general adoption of herbicides and mechanisation. Some changes in the weed flora have occurred usually associated with monocropping and the continued use of one herbicide. For example, chlorpropham used selectively in some crops has been responsible for increased populations of Pineappleweed (*Matricaria matricarioides*) and Gallant Soldier (*Galinsoga parviflora*), and where herbicides have enabled horticultural crops to be grown as part of a farm rotation, some predominantly agricultural weeds such as Wild-oat (*Avena fatua*) have occurred in peas, carrots and various horticultural brassicas. But, perhaps surprisingly, the major weeds occurring in vegetable crops in 1975 were almost the same as those recorded in 1953 (Davison and Roberts, 1976).

Another increasing trend is the local emergence of herbicide resistant forms in previously susceptible species. The first such report was from Hawaii (Hanson, 1959) where a 2,4-D resistant form of the broad-leaved sugar cane weed *Erechtites hieracifolia* was found. More recently repeated annual application of triazine herbicides has produced atrazine resistant Groundsel (*Senecio vulgaris*) and other species. Deliberate breeding and selection, too, has revealed considerable potential for resistance in crops and weeds (Chapter 6). Fortunately the emergence of herbicide resistance within weed species seems unlikely to pose problems similar to those created by build-up of pesticide resistance in insects and fungi (Harper, 1956). On the other hand the shift in dominance from broad-leaved weeds to grasses or sedges has created serious problems in Britain and elsewhere.

## REFERENCES


AMOR, R. L. and HARRIS, R. V. (1975). Seedling establishment and vegetative spread of *Cirsium arvense* (L.) Scop. in Victoria, Australia, *Weed Res.*, **15**, 407–411

BAKER, H. G. (1965). Characteristics and modes of origin of weeds, in *The Genetics of Colonizing Species* (ed. H. G. Baker and G. L. Stebbins), Academic Press, New York, p. 588

BAKER, H. G. (1972). Migration of weeds, in *Taxonomy, Phytogeography and Evolution* (ed. D. H. Valentine), Academic Press, London, p. 431

BAKER, H. G. (1974). The evolution of weeds, *Ann. Rev. Ecol. Syst.*, **5**, 1–24

BLACK, C. C., CHEN, T. M. and BROWN, R. H. (1969). Biochemical basis for plant competition, *Weed Sci.*, **17**, 338–344

BURROWS, F. M. (1974). Wind dispersal of weed seeds, *Proceedings 12th British Weed Control Conference*, pp. 1121–1129

DAVISON, J. G. and ROBERTS, H. A. (1976). Influence of changing husbandry on weeds and weed control in horticulture, *Proceedings 1976 British Crop Protection Conference – Weeds*, **2**, 1009–1017

FRYER, J. D. and CHANCELLOR, R. J. (1970). Herbicides and our changing weeds, in *The Flora of a Changing Britain*, B.S.B.I. report 11, London

GAUDET, J. J. (1979). Aquatic weeds in African man-made lakes, *PANS*, **25** (3), 279–286

GLOB, P. V. (1969). *The Bog People*, Faber and Faber, London, p. 142

GODWIN, H. (1960). The history of weeds in Britain, in *The Biology of Weeds* (ed. J. L. Harper), Blackwell, Oxford, p. 256

HANSON, N. S. (1959). Chemical weed control in Hawaii, *Proceedings 10th Congress of the International Society of Sugar Cane Technologists*, pp. 538–549
HARPER, J. L. (1956). The evolution of weeds in relation to resistance to herbicides, *Proceedings 3rd British Weed Control Conference*, pp. 176–188
HARPER, J. L. (1965). Establishment, aggression and cohabitation in weedy species, in *The Genetics of Colonizing Species* (ed. H. G. Baker and G. L. Stebbins), Academic Press, New York, p. 588
HASELER, W. H. (1976). *Parthenium hysterophorus L.* in Australia, *PANS*, **22**, 515–517
HELBAEK, H. (1953). Early crops in Southern England, *Proc. prehist. Soc.*, **18**, 194–224
HILL, T. A. (1977). *The Biology of Weeds*, Arnold, London, p. 64
HOLM, L. (1976). The importance of weeds in world food production, *Proceedings 1976 British Weed Control Conference – Weeds*, **3**, 753–769
HORNE, F. R. (1953). The significance of weed seeds in relation to crop production, *Proceedings 1st British Weed Control Conference*, pp. 372–398
HORNSEY, K. G. and ARNOLD, M. H. (1979). The origins of weed beet, *Ann. appl. Biol.*, **92**, 279–285
HUDLER, G. and FRENCH, D. W. (1976). Dispersal and survival of seed of eastern dwarf mistletoe, *Can. J. Forest Res.*, **6**, 335–340
KELLEY, A. and BRUNS, V. (1975). Dissemination of weed seeds by irrigation water, *Weed Sci.*, **23**, 486–493
KHOSLA, S. N. and SOBTI, S. N. (1979). *Parthenium* – a national health hazard, its control and utility – a review, *Pesticides*, **13** (7), 21–26
LONGDEN, P. C. (1976). Annual Beet: problems and prospects, *Pestic. Sci.*, **7**, 422–425
MITCHELL, D. S. (1972). The Kariba weed: *Salvinia molesta, Brit. Fern Gaz.*, **10**, 251–252
PIGGIN, C. M. (1978). Dispersal of *Echium plantagineum* by sheep, *Weed Res.*, **18**, 155–160
PRICE JONES, D. (1976). *Wild Oats in World Agriculture*, Agricultural Research Council, London, p. 296
SALISBURY, E. (1961). *Weeds and Aliens*, Collins, London, p. 384
STEBBINS, G. L. (1971). *Chromosomal Evolution in Higher Plants*, Arnold, London, p. 216

## FURTHER READING

GRIME, J. P. (1979). *Plant Strategies and Vegetation Processes*, Wiley, Chichester, p. 222
HARPER, J. L. (1977). *Population Biology of Plants*, Academic Press, New York, p. 892
HOLM, L. G., PLUCKNETT, D. L., PANCHO, J. V. and HERBERGER, J. P. (1977). *The World's Worst Weeds*, University Press of Hawaii, Honolulu, p. 609
KING, L. J. (1966). *Weeds of the World: Biology and Control*, Leonard Hill, London, p. 526
McNEIL, J. (1976). Taxonomy and evolution of weeds, *Weed Res.*, **16**, 399–413
STIRTON, C. H. (1978). *Plant Invaders – Beautiful but Dangerous*, Dept. of Nature and Environmental Conservation of the Cape Provincial Administration, Cape Town, p. 175 (includes drawings and coloured photographs of 26 alien weeds of South Africa)

# 3 ONE YEAR'S SEEDS, SEVEN YEARS' WEEDS

## 3.1 WHERE WEEDS COME FROM

Whenever fields are cultivated, weed seeds and propagules of perennial species germinate and grow. Most have been produced *in situ* in weedy crops of the past, a fact that has given rise to the prediction embodied in the title of this chapter. In the context of this book the term seed includes true seeds and the functional seeds (actually fruits) produced in the grasses and in several other plant families. Most perennial weeds produce seeds but may in addition reproduce asexually by means of bulbs, rhizomes, runners and other structures. Seed production by several species, including Dandelion (*Taraxacum officinale*) and Bramble (*Rubus fruticosus*) is essentially asexual because the seed is not always produced following sexual fusion, but instead generally contains diploid cells which are identical genetically with each other and with the parent. From the weed control point of view, such apomictic seeds are essentially similar to the homozygous ones produced by habitually self-pollinating species. In addition to increase by seed and vegetative means *in situ*, seeds are brought to the field as contaminants in crop seed, in soil, manure, straw or on farm machinery, in irrigation water and attached to animals; many seeds, especially those of the Compositae and of some trees, are windborne. Propagules arriving with the crop seed are in a specially advantageous position, often providing new introductions. For example, the recent spread of Wild-oats in Britain is thought to be due mainly to introductions with crop seed but also with straw, manure and from contaminated combine harvesters.

Simplified balance sheets of seed gains and losses and flux diagrams representing the dynamics of seed and adult plant populations have been produced (Mortimer, 1976). It is hoped that the information gained can be used when attempting to eliminate a particularly troublesome species by steadily reducing the numbers of its seeds in the soil, and by indicating the most vulnerable phases in the life cycle.

### 3.1.1 Seed production

A feature common to most kinds of weeds is rapid reproduction; and estimates of the seed production by single weed plants certainly make alarming reading (Salisbury, 1942; Stevens, 1957). Many species can produce several thousand seeds per plant, and there are records of over one hundred thousand per plant. Any species that is permitted to produce seed so prolifically may become locally dominant in succeeding crops in that field. Of greater practical significance than production by single isolated plants, is the total seed production per unit land area, and as many as 60,000 viable seeds produced per square metre per year are not uncommon in horticultural cropping in Britain. Fortunately, many of these seeds perish by being eaten by birds or other animals or are attacked by pathogenic fungi. Other seeds germinate, but the seedlings are destroyed by cultivation, thus preventing the return of weed seeds to the soil.

## 3.2 BURIED WEED SEEDS IN SOIL

### 3.2.1 Estimation of numbers of seeds buried in soils

Any method of analysis of the factors controlling weed seed numbers depends first on an accurate method of enumeration. Seeds buried in the soil soon become superficially eroded by microorganisms, making them hard to distinguish from other organic debris. A small proportion only of the seeds in a soil sample will germinate immediately, although exposure to light, to cycles of alternating high and low temperature, or to solutions of nitrates, ethephon or sodium azide, often increases germination (Chancellor *et al.*, 1971; Hall and Wareing, 1972; Fay and Gorecki, 1978). A statistically valid soil sampling technique is also essential because of the great variations normally found within short distances in both total numbers of seeds and in the relative proportions of different weed species. The method now generally used for determining the viable seed population is that described by Brenchley and Warrington (1930) which has since been extensively used by Roberts (1970). In this method moist soil from field samples is exposed in shallow clay pans in a cool greenhouse; after removal and counting of each flush of seedlings the soil is stirred to encourage further germination. After two years, or when seedling emergence has become infrequent, the total numbers of each species per square metre of the sample plot is estimated from the counts recorded from the pans.

Using the open pan method, values from 2,000 seeds/m$^2$ to over 86,000/m$^2$ of topsoil have been claimed, Roberts and his coworkers finding 10,000/m$^2$ to be the median number of viable seeds in the upper 15 cm of the intensive vegetable production areas that they sampled. However, it is generally conceded that the open pan method underestimates the actual numbers present because many viable seeds probably fail to germinate within two years; Hurle (1974), who used the same method for agricultural soils in Germany, kept his pans for four years and exposed the soil samples to freezing temperatures to enhance germination, but still considered that numbers of some species were underestimated.

Other methods of assessing weed seed numbers rely on flotation of the organic soil fraction on a heavy liquid (such as a solution of potassium carbonate with a density of 1.56) followed by sieving and hand separation of the seeds and then testing for seed viability by germination (Kropac, 1966). Although a similar method is successful when counting nematode cysts in soil, it has not been adopted widely for weed seed estimation. A new method of seed separation using air flotation of the organic debris left after sieving a soil slurry has been described by Standifer (1980).

### 3.2.2 The dynamics of weed seed populations in the soil

The record for retention of viability is claimed for Arctic Lupin, seeds of which were recovered from animal burrows in the frozen tundra of Northern Canada, where it is thought they must have lain undisturbed for over 10,000 years (Porsild *et al.*, 1967). Odum (1974 and 1978) produced evidence that many common European weed seeds have survived in soil for at least 100 years and that a few species, including Fat-hen (*Chenopodium album*), may retain their viability in soil for over 600 years. In Britain, land that has been continuously under grass within living memory has, when ploughed, grown weeds typical of an earlier period under arable cropping. Researchers in Europe and North America have buried crop and weed seeds in soil under controlled conditions, recovering samples at intervals to assess the fall in viability, and although such contrived situations probably have little practical relevance, they have all demonstrated that buried weed seeds generally last considerably longer than crop seeds. Recently, Roberts and coworkers (1970) have shown that the turnover of weed seeds in soil is surprisingly rapid, and that an undisturbed soil may lose between 20 and 30 per cent of its viable seed population within a year; cultivation increases the decline to about 40 per cent and with very frequent cultivation to as much as 56 per cent (*Figure 3.1*). The decline in weed seed numbers follows an exponential decay curve, analagous to the decay observed in radioactive elements, so that a 50 per cent decline within one year corresponds to a half-life of one year. When all seed inputs were prevented a consistent percentage decline occurred each year over a five-year period. The rates of decline for individual species were shown to vary, with Common Chickweed (*Stellaria media*) and Annual Meadow-grass (*Poa annua*) losing about half their viable seeds per year, while the rate of loss of Fumitory (*Fumaria officinalis*) was less than 30 per cent. It might be expected that large disparities in the rates of decline between different species would eventually cause differences in the relative abundance of their seeds and as a result change the shape of the overall decay curve but

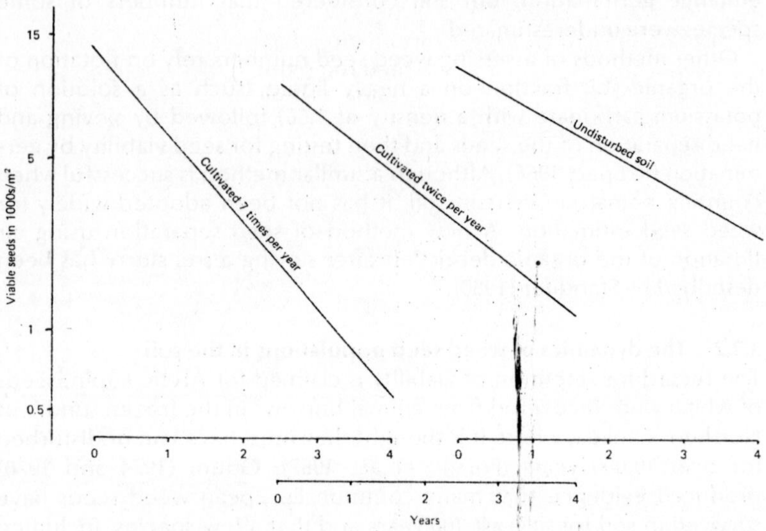


**Figure 3.1** Decrease in numbers of viable seeds in upper 23 cm of soil (15 cm for cropped soil) in absence of further seed production. (From Roberts and Feast, 1973.)

this was not noticed during the five year duration of the experiments.

One overwhelming fact that emerges from these studies is the sheer size of the seed reservoir, especially under intensive systems of farming that provide many opportunities for weed reproduction. Beginning with a population of 10,000/m² and assuming, optimistically, a sustained 50 per cent annual decline in numbers attained by frequent cultivation, this large population could be reduced after five years to about 600/m². Assuming that about 5 per cent of these 600 produced seedlings after cultivation, an assumption borne out by experiments (Roberts and Ricketts, 1979), then some 30 plants/m² would be available to compete with a crop; left uncontrolled such a population could, in a single season, increase the weed seed reserve to its former levels (Hurle, 1974). Complete prevention of seed production for such a long period is in any case impracticable, although it is possible to work towards the eradication of one particular species, such as Wild-oat (*Avena fatua*), when that species shows a reasonably rapid decline in seed numbers (Wilson, 1978). Within the last century two formerly common cornfield weeds, Darnel (*Lolium temulentum*) and Corncockle (*Agrostemma githago*), disappeared from fields in Britain because their buried seeds lost viability quickly and effective seed cleaning prevented regular reintroduction (Chapter 2).

Cultivation has the effect of stimulating weed germination, in part at least because it exposes seeds to light (Wesson and Wareing, 1969a,b; Van Rooden *et al.*, 1970), and germination due to cultivation is perhaps the largest single cause of viable seed reduction. Most weed seeds germinate near the soil surface, coming mainly from the upper 5 cm, although the seedlings of a few species with large seeds can emerge from the full depth of cultivation. It is probable that many weed seeds die because they are stimulated to germinate during cultivation only to be buried again too deeply for their shoots to reach the surface before food reserves in the seed have been exhausted. Other sources of loss are the actions of microorganisms and of seed eating animals. The reserve of viable weed seeds remaining at greater depths in the soil, which fails to be stimulated to germinate when reduced cultivation is practised, acts as a powerful buffer which moderates changes such as selection within or between species for herbicide resistance.

### 3.2.3 Dormancy and germination of weed seeds

Crop seeds germinate readily when sown into a good seedbed that is sufficiently warm and moist, but only 4 or 5 per cent of the viable weed seeds in the soil do so following seedbed preparation. This pattern of discontinuous germination seen in most successful weeds is common also among wild plants in general, acting as a survival mechanism preventing simultaneous germination of all the seeds and only encouraging it when reproduction can be assured. Several distinct patterns of climatic adaptation among common British weeds are illustrated in Fryer and Makepeace (1977). Predominantly cool-season germinators, such as Winter Wild-oat (*Avena ludoviciana*), Black-grass (*Alopecurus myosuroides*), Cleavers (*Galium aparine*) and Ivy-leafed Speedwell (*Veronica hederifolia*), are more troublesome in autumn sown crops that stand over winter, while spring and early summer germinators will be the major problem in spring sown crops. Germinability of Knotgrass (*Polygonum aviculare*) is promoted by winter chilling, germination then occurring as the soil warms in the spring; but as the temperature continues to rise in early summer, the chilling effect is reversed and remaining seeds in the soil become dormant again. Distinctly different climatically controlled patterns occur in other species (Roberts and Lockett, 1978); however, ephemeral species that produce several generations each year, such as Annual Meadow-grass (*Poa annua*), Common Field-speedwell (*Veronica persica*) and Common Chickweed (*Stellaria media*), germinate well at all seasons provided conditions are not too severe.

The term dormancy can be used in a general way to mean lack of activity or growth, but a more specific meaning is usually implied, its use being restricted to situations where a seed or bud fails to grow despite reasonable conditions for growth, such as warmth, moisture and a supply of oxygen (Roberts, 1972). Dormant seeds may be unable to grow for several different reasons, acting either alone or in combination. Some plants, including most orchids, produce extremely small seeds which require a prolonged period of development before the seedling can emerge above ground, but this form of dormancy is rare in weeds. Freshly shed seed is seldom ready for immediate germination, requiring first a period of so-called after ripening; presumably this has evolved to prevent germination occurring before dispersal or before the beginning of the next growing season. Many seeds are enclosed in hard or impermeable outer layers, derived from either seed or fruit, which prevent water entry or gaseous interchange, or sometimes physically prevent enlargement of the embryo. Other seeds contain chemical inhibitors which must be washed away, perhaps after erosion of an enclosing waterproof seed coat, before germination can proceed; the cultivated beets retain this feature and allowance must be made for it when testing beet seed for viability. Another common requirement of seeds from the temperate regions is for a period of winter chilling before dormancy is overcome as in the case of Knotgrass (*Polygonum aviculare*) already described. Some kinds of dry stored seeds germinate fully only after exposure to light of the imbibed seeds, the best known example being seed of lettuce cultivar Grand Rapids. Because the dry seeds of the majority of species germinate as well in the dark as in the light, a light requirement was not regarded as important in relation to weed seeds buried in the soil. Most seeds survive in the soil in the imbibed state, however, rather than dry, as in a seed packet, and it has been shown (Wesson and Wareing, 1969*b*) that imbibed weed seeds that are prevented from early germination by being buried too deeply may have an absolute requirement for light as a prelude to germination. Light requirement is known to be an example of the reversible red/far-red reaction involving the pigment phytochrome, in which red light of wavelength 660 nm is stimulatory and far-red light of 735 nm is inhibitory, with daylight acting essentially as red light (Taylorson, 1972). Green light that has filtered through a foliate canopy has lost most of its red component and fails to stimulate germination, which has the effect of suppressing weed emergence when the shade is too heavy for seedlings to survive to maturity.

The optimum temperature for germination may be only a little above freezing for some winter germinating species, in which high summer temperatures are often inhibitory. The concept of optimum temperature is hard to apply to weed seeds because most do not respond best to any one uniform temperature but germinate most readily when exposed to a daily alternation of low night temperatures and warm daytime ones. Such a diurnal variation is sometimes sufficient to promote germination in dormant seeds that, at a uniform temperature, would have a light requirement; such complex and, at first sight, baffling interactions between the factors that affect dormancy are the rule rather than the exception. The delicacy of the response of seeds to temperature is remarkable; for example, the dormancy of Common Poppy (*Papaver rhoeas*) seed from near Oxford is best broken by a daily alternation between 10°C and 30°C, which corresponds to the expected diurnal variation in the surface layers of the soil in April–May at the time when this weed germinates locally in cereal crops (Harper, 1957). The relationship between temperature and germination appears to be genetically stable for each species, and to be little affected by selection. It provides valuable information on the geographic origin of both crops and weeds (Thompson, 1971).

### 3.2.4 Types of dormancy in weed seeds

Dormancy in seeds results from the interaction of internal factors with environmental ones, and a confusing nomenclature of the various levels and kinds of dormancy has arisen. This was simplified by Harper (1957) who suggested that seeds are either born dormant, become dormant or else have dormancy thrust upon them (*Figure 3.2*). Those 'born dormant' possess innate dormancy, which remains in force until the seed experiences sufficient after ripening, winter chilling, erosion of the seed coat, or leaching of any chemical inhibitors present in the seed or seed coat. Under suitable conditions, seed released from innate dormancy may then germinate, but if this is prevented by, for example, the seed being buried too deeply where oxygen levels are low and carbon dioxide levels relatively high, then a state of deep dormancy may be induced, which will persist long after removal of the constraint. Seeds 'made dormant' in this way behave much as do those with innate dormancy. The third category, seeds that have dormancy 'thrust upon them', a condition called enforced dormancy, have sufficient water and are in almost all other respects ready for germination, only failing to do so because of some missing factor, which is frequently exposure to light or to alternating high and low temperatures. Seeds appear to be predisposed to enforced dormancy by prolonged exposure to high levels of carbon dioxide, and possibly to gaseous inhibitors produced by the seed itself, to darkness and to a uniform temperature, all of which will be experienced by seeds that have been buried deeply during cultivation or by worms. Unlike seeds under induced dormancy, these seeds germinate readily when the missing factor is supplied, provided that the other conditions are suitable. Seeds brought to the surface during cultivation will be exposed to light, to fluctuating temperatures and possibly to raised levels of nitrate, any of which alone, and especially when combined, can remove the restraints of dormancy. For example, Taylorson and

McWhorter (1969) found that individually all three factors slightly raised the percentage germination of Johnson Grass (*Sorghum halepense*) but that the three together caused substantial increases; similar results have been obtained with several kinds of annual weeds in Britain, a notable exception being Corn Spurrey (*Spergula arvensis*) (Vincent and Roberts, 1977). Differences may also occur depending on the age of the seed; old seed of Fat-hen (*Chenopodium album*)

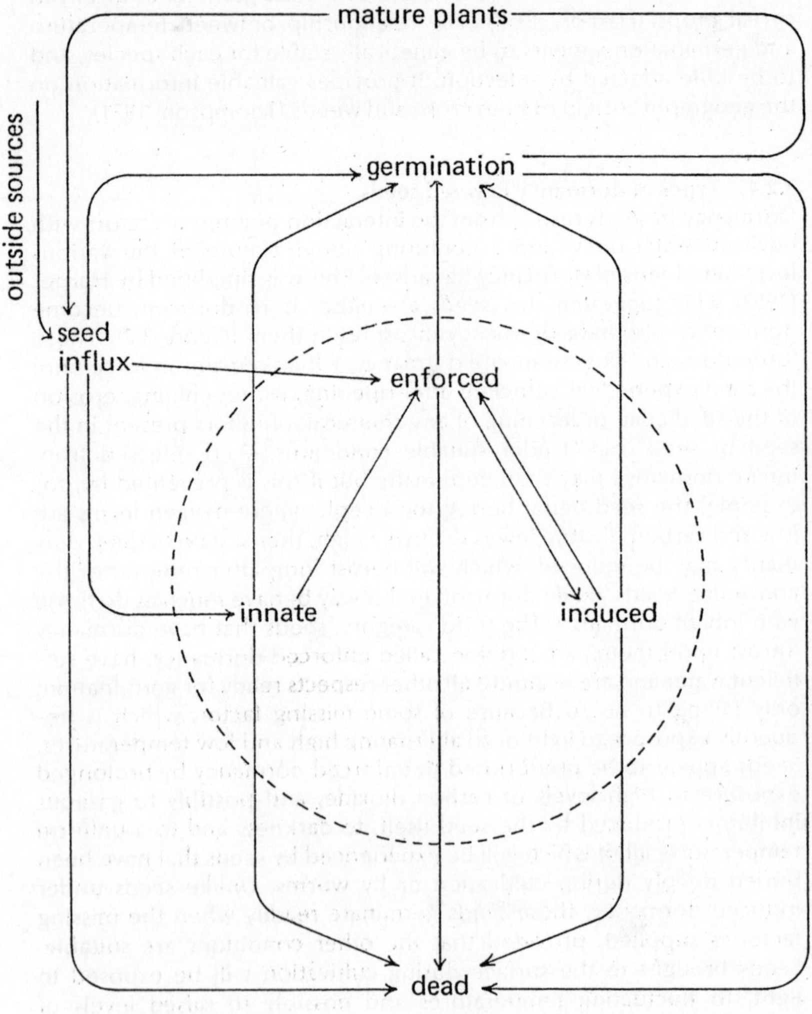


**Figure 3.2** Scheme to show the dynamics of the soil population of weed seeds. The soil is indicated as a broken line circle. (Devised by Roberts, 1970.)

responded with increased germination to light or to nitrate solution, while younger seed required both simultaneously. Nitrate levels sufficiently high to stimulate germination may occur near the soil surface especially after addition of fertiliser (Fawcett and Slife, 1978).

The interrelations of the different dormancy states and their relation to the overall dynamics of the weed seed population in the soil are shown in *Figure 3.2*. The duration of innate dormancy within the progeny of one plant varies, and often follows a normal distribution. Occasional and irregular flushes of germination may suggest another type of distribution; but there can be other explanations. For example, this behaviour in Wild-oat (*Avena fatua*) has been attributed to seasonal periods of great microbial activity leading to breakdown of the pericarp, which then allows chemical inhibitors to escape, thus releasing the seed from innate dormancy.

### 3.2.5 Weed seed polymorphism

There are many examples of seed polymorphism in which visible differences in colour, size or form are matched by behavioural ones. In the Cocklebur (*Xanthium pennsylvanicum*) the fruits contain two seeds of unequal size, the larger one generally germinating a year before the smaller. Although the more prolonged dormancy of the smaller seed has been linked with its thicker and more impervious testa, this may be of secondary importance to the presence of a chemical germination inhibitor. In Fat-hen (*Chenopodium album*) there are four visibly different types of seed each with characteristic dormancy behaviour (Williams and Harper, 1965). Generally these seed polymorphs are not susceptible to selection, being produced in the same proportions irrespective of the seed type of the parents; but the environment of the parent plant can profoundly influence some types of polymorphism, and under long days Halogeton (*Halogeton glomeratus*) produces brown seeds with prolonged innate dormancy, but less dormant black seeds under short days. Short days also increase dormancy in Fat-hen. In Corn Spurrey (*Spergula arvensis*), seed polymorphism is not only linked to physiological responses of the seed but of the whole plant, and the relative proportions of smooth and knobbly seeds in a population have been shown to vary with altitude and latitude (New, 1958).

A physiological polymorphism, without obvious physical difference, often occurs, which gives rise to peaks in germination caused by abrupt discontinuities in the distribution of innate dormancy duration. This is generally seen when comparing the progeny of the same species grown in differing habitats, similar to the daylength effect already mentioned, but is also common between adjacent and apparently identical plants of the same species; it can also be seen in some weeds according to the position of the seed on the inflorescence (Cavers and

Harper, 1966). Together with variability in the subsequent imposition of induced or enforced dormancy, the intermittent germination of weed seeds over a long period is assured.

### 3.2.6 Seeds of parasitic weeds

Seeds of certain root parasites, including *Orobanche* and *Striga* spp., are innately dormant, having an absolute requirement for chemical stimulation by a susceptible host plant before germination can proceed (Edwards *et al.*, 1976). Several synthetic organic substances stimulate some *Striga* spp. to germinate in the absence of a suitable host plant (Johnson *et al.*, 1976), but so far the treatment is too expensive as a method of control. A similar technique has been tested for inducing common British weeds to germinate, using the ethylene generator ethephon but here, too, considerable practical problems remain to be solved (Chancellor *et al.*, 1971).

## 3.3 PROPAGATION AND SURVIVAL OF PERENNIAL WEEDS

Some of the weeds that cause most concern are the perennial species with an effective means of vegetative reproduction, especially those that seem to thrive on cultivation. Weeds arising from seeds, because they are more susceptible to well-timed cultivations, and to herbicides, are relatively easy to control. The perennating organs of perennial weeds can be stolons, runners, rhizomes, tubers of various kinds, bulbs and bulbils and roots. All too often the cultivations or herbicides that are effective in controlling weed seedlings only make infestation of perennials worse by cutting up rhizomes and roots or scattering bulbils (*Table 3.1*). In some species of perennial weeds, reproduction is almost entirely vegetative; for example, the Slender Speedwell (*Veronica filiformis*) or the Horse-radish (*Armoracia rusticana*) normally do not set seed in Britain. In other weeds seeds form an important part of the life history, and are the usual means of entry to a new situation, although spread thereafter may well be mainly vegetative. In the newly drained polders of Holland, infestations of the invasive perennial weeds Creeping Thistle (*Cirsium arvense*) and of Coltsfoot (*Tussilago farfara*) were initiated by seed but were then sustained by vigorous vegetative reproduction.

Among the most troublesome perennial weeds are the rhizomatous and stoloniferous grasses, including Common Couch (*Agropyron repens*), Bermuda-grass (*Cynodon dactylon*), Johnson-grass (*Sorghum halepense*) and Lalang (*Imperata cylindrica*), all of which can be spread by seed and vegetatively; Common Couch, for example, has been shown to be capable of regeneration from rhizome fragments no more than 1.25 mm long. Nut forming sedges, especially Yellow Nutsedge (*Cyperus esculentus*) and Purple Nutsedge (*C. rotundus*) are

*Table 3.1*

*Vegetative regeneration methods available to weeds*

| | *Example* |
|---|---|
| 1. Creeping, generally horizontal stems | |
| rhizomes – in the soil | *Agropyron repens* |
| tubers – generally in the soil | *Cyperus rotundus* |
| stolons – above the soil | *Ranunculus repens* |
| 2. Swollen shoot bases – at or below soil surface | |
| bulbs | *Allium* and *Oxalis* spp. |
| 3. Independent growth of detached pieces of stem | |
| turions | Winter resting phase of several aquatic weeds |
| broken stem pieces | Willows in New Zealand; *Veronica filiformis* |
| 4. Specialised roots | |
| root suckers | *Cirsium arvense* |
| tap roots – broken up in cultivation | *Rumex obtusifolius* |
| 5. Non-sexual seed formation | |
| apomictic seed production | *Taraxacum officinale* |

amongst the worst of all weeds in warmer regions throughout the world because the tubers and rhizomes are formed so prolifically. As many as 50,000 nuts per cubic metre of soil have been recorded, each of which can produce a new plant (Kasasian, 1971). Similar problems are posed in some areas by the bulb-forming weed species of *Oxalis*, especially the so-called Bermuda Buttercup (*Oxalis pes-caprae*) from South America. In temperate regions the several kinds of docks (*Rumex* spp.), thistles (mainly *Cirsium* spp.), and Field Bindweed (*Convolvulus arvensis*) also withstand cultivation well, but like most perennial weeds they also thrive under reduced cultivation regimes in which both soil disturbance and annual weed competition are lacking.

Several perennial weeds, including Creeping Thistle, Field Bindweed and Horsetails (*Equisetum* spp.) possess substantial vegetative structures buried below the depth of cultivation from which new shoots arise to replace those killed by cultivation or by the action of herbicides. Complete eradication of such weeds prior to planting is therefore virtually impossible.

### 3.3.1 Introduction of perennial weeds

The importance of perennial weed introduction by contaminated crop seed is widely recognised, and in Britain there is a legal requirement (Anon, 1961; Mackay, 1964) that the suppliers of cereal and herbage should declare the weight, as a percentage, of named noxious weeds. The list includes the annual species Black-grass (*Alopecurus myosuroides*), the Wild-oats (*Avena* spp.) and Dodder (*Cuscuta epithymum*), the first two of which are of major importance in cereal crops. In addition, the list includes the perennial Common Couch (*Agropyron repen*) and the Docks (*Rumex* spp.). Common Couch is self-sterile and seed production from any one field may be infrequent if only a single clone is present; introduction of new clones via contaminated crop seed is therefore likely to increase the percentage of viable seeds. The occurrence of such a restriction in seed set by the number of clones present seems likely in view of the observed variation from 10 per cent to over 60 per cent in set of caryopses (MacKay, 1964). In North America the spread of Johnson-grass is also attributed to the use of contaminated crop seeds.

## 3.4 ONE YEAR'S SEEDS, SEVEN YEARS' WEEDS?

Weed infestations arise from the seeds and perennating organs already present in the soil and from propagules newly introduced from outside. The latter, although quantitatively insignificant, may be of outstanding importance qualitatively when a noxious new weed is introduced. The work of Roberts and others has shown that seeds, on average, have a rather short residence time in the soil, although individual seeds of some species may last for centuries. This encourages the belief that a sustained and effective weed control programme could lead to the local eradication of some of the more unwelcome species, such as the Wild Oats in Britain and Witchweeds in parts of North America.

On an immediately practical level, cultural techniques that reduce the return of weed propagules to the soil, that encourage weed seed germination followed by the destruction of seedlings, and keep freshly fallen seeds near the surface where losses are greatest, will all reduce the potential for future weed problems.

## REFERENCES

ANON (1961). *The Seeds Regulations*, HMSO, London

BRENCHLEY, W. E. and WARRINGTON, K. (1930). The weed seed populations of arable soil. 1. Numerical estimations of viable seeds and observations on their natural dormancy, *J. Ecol.*, **18**, 235–272

CAVERS, P. B. and HARPER, J. L. (1966). Germination polymorphism in *Rumex crispus* and *Rumex obtusifolius*, *J. Ecol.*, **54**, 367–382

CHANCELLOR, R. J., PARKER, C. and TEFEREDEGN, Taye (1971). Stimulation of dormant weed seed germination by 2-chloroethylphosphonic acid, *Pestic. Sci.*, **2**, 35–37

EDWARDS, W. G., HIRON, R. P. and MALLET, A. I. (1976). Aspects of the germination of *Orobanche crenata* seed, *Z. Planzenphysiol. Bd.*, **80**, 105–111

FAWCETT, R. S. and SLIFE, F. W. (1978). Effects of field applications of nitrate on weed seed germination and dormancy, *Weed Sci.*, **26**, 594–596

FAY, P. K. and GORECKI, R. S. (1978). Stimulating germination of dormant wild oat (*Avena fatua*) seed with sodium azide, *Weed Sci.*, **26**, 323–326

FRYER, J. D. and MAKEPEACE, R. J., *Weed Control Handbook*, vol. I: *Principles*, Blackwells, Oxford, 510 pp.

HALL, M. A. and WAREING, P. F. (1972). Dormancy in weed species, *Proceedings 11th British Weed Control Conference*, pp. 1173–1182

HARPER, J. L. (1957). The ecological significance of dormancy and its importance in weed control, *Proceedings 4th International Congress on Plant Protection*, Hamburg, pp. 415–420

HURLE, K. (1974). Effects of long-term weed control measures on viable weed seeds in the soil, *Proceedings 12th British Weed Control Conference*, pp. 1145–1152

JOHNSON, A. W., ROSEBERRY, G. and PARKER, C. (1976). A novel approach to *Striga* and *Orobanche* control using synthetic germination stimulants, *Weed Res.*, **16**, 223–227

KASASIAN, L. (1971). *Weed Control in the Tropics*, Leonard Hill, London, 307 pp.

KROPAC, Z. (1966). Estimation of weed seeds in arable soil, *Pedobiologica, Bd.*, **6**, 105–128

MACKAY, D. B. (1964). The incidence and significance of injurious weed seed in crop seed, *Proceedings 7th British Weed Control Conference*, pp. 583–591

MORTIMER, A. M. (1976). Aspects of the seed population dynamics of *Dactylis glomerata* L., *Holcus lanatus*, *Plantago lanceolata*, and *Poa annua* L., *Proceedings 1976 British Crop Protection Conference – Weeds*, pp. 687–694

NEW, Jane K. (1958). A population study of *Spergula arvensis*. 1. Two clines and their significance, *Annls Bot. N.S.*, **22**, 457–477

ODUM, S. (1974). Seeds in ruderal soils, their longevity and contribution to the flora of disturbed ground in Denmark, *Proceedings 12th British Weed Control Conference*, pp. 1131–1144

ODUM, S. (1978). *Dormant Seeds in Danish Ruderal Soils*, Royal Veterinary and Agricultural University, Horsholm, Denmark, 247 pp.

PORSILD, A. E., HARRINGTON, C. R. and MULLIGAN, G. A. (1967). *Lupinus arcticus* grown from seeds of Pleistocene age, *Science*, **158**, 113–114

ROBERTS, E. H. (1972). Dormancy: a factor affecting seed survival in the soil, in *Viability of Seeds* (ed. E. H. Roberts), Chapman and Hall, London, p. 448

ROBERTS, H. A. (1970). Viable weed seeds in cultivated soils, *Rep. natn. Veg. Res. Stn., Wellesbourne, 1969*, 25–38

ROBERTS, H. A. and FEAST, P. M. (1973). Changes in the numbers of viable seeds in soil under different regimes, *Weed Res.*, **13**, 298–303

ROBERTS, H. A. and LOCKETT, Patricia M. (1978). Seed dormancy and field emergence in *Solanum nigrum*, *Weed Res.*, **18**, 231–241

ROBERTS, H. A. and RICKETTS, Margaret E. (1979). Quantitative relationships between the weed flora after cultivation and the seed population in the soil, *Weed Res.*, **19**, 269–275

SALISBURY, E. (1942). *The Reproductive Capacity of Plants*, Bell, London

STANDIFER, L. C. (1980). A technique for estimating weed seed populations in cultivated soil, *Weed Sci.*, **28**, 134–138

STEVENS, O. A. (1957). Weights of seeds and numbers per plant, *Weeds*, **5** (1), 46–55

TAYLORSON, R. B. (1972). Phytochrome controlled changes in dormancy and germination of buried weed seeds, *Weed Sci.*, **20**, 417–422

TAYLORSON, R. B. and McWHORTER, C.. (1969). Seed dormancy and germination in ecotypes of Johnson-grass, *Weed Sci.*, **17**, 359–361

THOMPSON, P. A. (1971). Research into seed dormancy and germination, *Proceedings International Plant Propagators Society Annual Meeting*, 211–228

VAN ROODEN, J., AKKERMANS, L. M. A. and VAN DER VEEN, R. (1970). A study of photoblastism in seeds of some tropical weeds, *Acta Bot. Neerl.*, **19**, 257–264

VINCENT, Elizabeth M. and ROBERTS, E. H. (1977). The interaction of light, nitrate and alternating temperature in promoting the germination of dormant seeds of common weed species, *Seed Sci. Technol.*, **5**, 1659–1670

WESSON, G. and WAREING, P. F. (1969*a*). The role of light in the germination of naturally occurring populations of buried weed seeds, *J. Exp. Bot.*, **20**, 402–413

WESSON, G. and WAREING, P. F. (1969*b*). The induction of light sensitivity in weed seeds by burial, *J. Exp. Bot.*, **20**, 414–425

WILLIAMS, J. T. and HARPER, J. L. (1965). Seed polymorphism and germination. 1. The influence of nitrates and low temperatures on the germination of *Chenopodium album*, *Weed Res.*, **5**, 141–150

WILSON, B. J. (1978). The long term decline of a population of *Avena fatua* L. with different cultivations associated with spring barley cropping, *Weed Res.*, **18**, 25–31

# 4 SPECIAL WEED PROBLEMS

Each kind of weed problem outlined in this chapter is capable of dominating the local economic and social life wherever it occurs. The recent rapid spread and explosive growth of aquatic weeds throughout the wet tropics has produced locally disastrous effects; in the temperate world the harm done by water plants is less dramatic but there has been increased weed growth in polluted water enriched with nutrients. Poisonous plants continue to take a heavy toll of livestock, and man himself may become ill or, less commonly today, actually die, when poisonous plants are eaten inadvertently. In addition, there are plants which although not lethal can in various ways have debilitating effects on man and his animals. Parasitic weeds, once a major concern of farmers in Northern Europe, have largely disappeared since seed cleaning and legislation have reduced the sowing of contaminated crop seed. However, in other areas root and stem parasites can occur on almost every major crop, posing severe local problems. Woody species have been taken by man to new areas, where they have sometimes escaped and now dominate the local vegetation; for example, the attractive shrubby Lantana (*Lantana camara*), which is a non-weedy native of tropical America, was planted by man in other tropical areas, where its vigorous growth, rapid increase and toxicity to livestock have made it a widespread nuisance.

## 4.1 AQUATIC WEEDS

These are amongst the most intractable of all weed problems, particularly in the tropics and subtropics. The presence of water plants confers some advantages, such as giving stability to the banks of watercourses, increasing the amount of dissolved oxygen in the water and providing food and shelter for animals, especially for fish and water fowl. However, in excess, water plants become weeds because of the problems they create. These include restriction of water flow, increased loss of water by evaporation, more rapid deposition of silt and organic matter which reduces the effective water depth, restriction of free navigation by boats and harbouring organisms involved in diseases of man and livestock. Some of the plants causing these problems, although rooted on the bed of the stream or in the banks, have their aerial parts emergent above the water surface, while other species remain submerged. Other kinds of weed are rooted in the stream-bed but have leaves that are not emergent or submerged, but instead float on the surface. Another group are fully free floating, where they can drift in currents or be blown by the wind into floating islands on which, in time, sufficient debris may accumulate to enable wild plants or sometimes useful crops, to take root (*Figure 4.1* and *Table 4.1*). Algae have a vital role as primary producers in aquatic food chains, but excessive seasonal growths or blooms can cause problems with piped drinking water by blocking filters and producing taint; blooms of some blue–green algae are also acutely toxic, killing fish and other organisms. When algal blooms and other weed growths decay, water can be temporarily deoxygenated thereby killing fish. Larger filamentous algae also form dense entangled growths known as 'blanket weed' in slow moving, nutrient-rich waters (Hawkins, 1972).

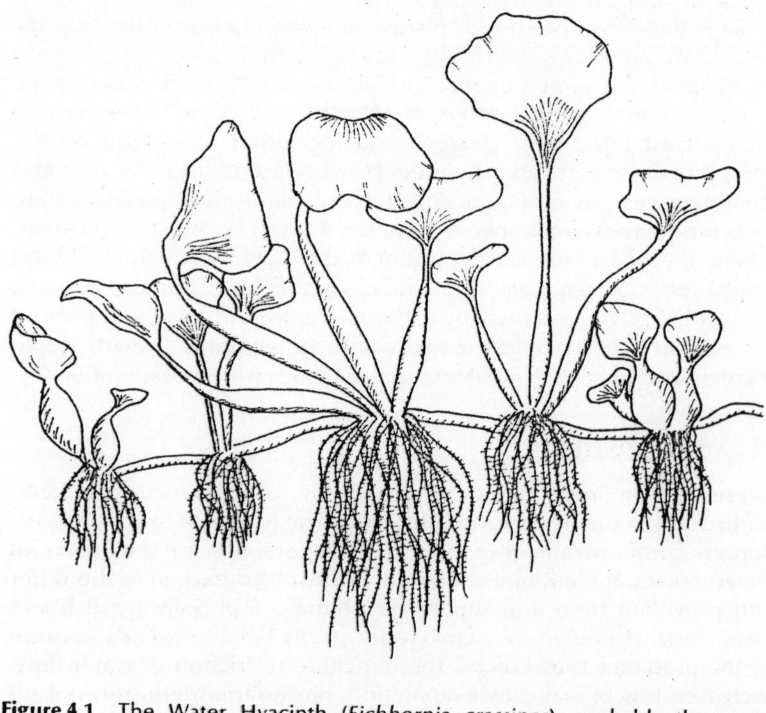

**Figure 4.1** The Water Hyacinth (*Eichhornia crassipes*), probably the most notorious of all water weed species, which has been spread throughout the tropics and subtropics. The beautiful blue flower spikes are not shown. (Water hyacinth plants can grow to a metre or more in height.)

*Table 4.1*

*Representative water weed species of various types from the tropics and temperate regions*

| | |
|---|---|
| 1. Emergent species that root in the bottom mud or the banks of watercourses whose foliage rises above the water. | |
| Common Reed | *Phragmites communis* |
| Great Reedmace | *Typha latifolia* |
| Reed-grass | *Glyceria maxima* |
| Papyrus and other sedges | *Cyperus papyrus* and other *Cyperus* spp. |
| 2. Plants rooted in the stream bed whose leaves float on the surface. | |
| Broad-leaved Pondweed | *Potamogeton natans* |
| Yellow Water-lily | *Nuphar lutea* |
| 3. Free-floating plants including several kinds of water fern. | |
| Duckweed | *Lemna minor* |
| Kariba Weed (= African Pyle) | *Salvinia molesta* |
| Water Hyacinth | *Eichhornia crassipes* |
| Water Lettuce | *Pistia stratiotes* |
| 4. Submerged or submersed plants, which are less obvious than the other groups but which collectively often cause the most severe water weed problems by reducing the rate of water flow. Can be rooted in the mud or free in the water. | |
| Canadian Pondweed | *Elodea canadensis* |
| Hydrilla | *Hydrilla verticillata* |
| Marestail | *Hippuris vulgaris* |
| Spiked (= Eurasian) Milfoil | *Myriophyllum spicatum* |
| 5. Algae forming 'cott' or blanket weed. | |
| | *Cladophora glomerata* |
| | *Vaucheria dichotoma* |

Most of the species causing problems as water weeds are perennials which continue growth throughout the year when conditions allow, but in temperate regions the aerial parts usually die back annually because of frost sensitivity. Overwintering of many species is by buds, called turions, which separate from the plant and can become dispersed before falling to the bottom mud. In the spring the turions resume growth. In some temperate and tropical water plants seed production is rare or at least is not the principal method of reproduction. For example, Water Hyacinth (*Eichhornia crassipes*) whose populations exist in the form of distinct clones, each of which tends to be self-incompatible, produces few seeds wherever a single clone predominates; Kariba Weed (*Salvinia molesta*), which is probably a sterile but vigorous hybrid between two non-aggressive species of South American water ferns, is apparently unable to produce spores (Mitchell, 1973).

The growth of water plants, especially the free-floating kinds, is particularly prolific in polluted waters containing high levels of phosphorus and nitrogen. Other factors that control the species composition and rate of growth include the water depth and rate of flow, the availability of light to submerged species and above all the suitability of the bottom mud and silt for weed growth (Fryer and Makepeace, 1977; Gaudet, 1979).

### 4.1.1 Economic losses due to aquatic weeds

Agricultural systems that rely on surface irrigation and an efficient system of drainage channels are very vulnerable to water weeds, which not only choke the channels, thereby reducing the rate of flow, but increase evaporation losses. In Florida the evapotranspiration from an area covered by Water Hyacinth (*E. crassipes*) can be four times as great as that from an equivalent open water surface (Timmer and Weldon, 1967); in India even greater rates of water loss occur. Seeds from weed-covered banks fall into irrigation water and thereby add to the seeds already in the soil, and although the total numbers are small compared with those produced *in situ* in weedy fields, the addition becomes important when the seeds are of species new to the area (Kelley and Bruns, 1975).

Dense mats of floating weeds interfere with gaseous exchange, reducing the oxygen dissolved in water and causing death of fish. This was a cause of disappointment in Lake Kariba, formed by damming the Zambesi, in which it had been hoped to establish commercial fishing. The first mats of Kariba Weed (*S. molesta*) were reported in 1959 and by 1962 over 10 per cent of the water surface was covered with the water fern and with Water Lettuce (*Pistia stratiotes*) (Gaudet, 1979). *Salvinia molesta* has also caused problems in irrigated rice, especially in Sri Lanka and India, and rice production has been abandoned in some of the worst affected areas (Thomas, 1979).

Water weeds harbour mosquitoes carrying malaria and other diseases, and a water snail that lives on weeds carries the organism responsible for bilharzia in man. Water Lettuce (*P. stratiotes*) has formed dense mats on the recently created Lake Volta in Ghana, where it is the preferred site for larvae of the mosquito vectors of encephalomyelitis and rural filariasis in man. One kind of mosquito, a species of *Mansonia*, actually remains submerged by obtaining its oxygen supply directly from the Water Lettuce, and this mosquito can only be controlled by destroying the host plants (Holm and Yeo, 1979).

Floating weeds can be spectacular and frightening in their effects, and are of great economic importance particularly in developing countries (Anon, 1971). Equally damaging are the plants rooted in the banks and the bottom mud, which reduce water-flow rates substantially. For example, Spiked Water-milfoil (*Myriophyllum spicatum*) has recently become a major pest in North America, where it is known locally as Eurasian Milfoil. It grows vigorously and has spread also to estuarine waters where fishing has been reduced and oyster beds smothered, but in addition reduces flow rates in fresh water channels. In India, bottom-rooted submerged species have in some cases reduced water flow in irrigation channels to a mere 20 per cent of the intended rate.

### 4.1.2 Control of water weeds

In the past control was by hand cutting and removal of weeds but much of the work is now mechanised. Herbicides can be used successfully, but the areas that must be sprayed in large lakes and rivers are immense, and because water serves so many purposes any herbicide applied must have a short persistence and an acceptably low toxicity to water animals, to crop plants and to man; the application of herbicides on or near watercourses is therefore subject to restrictions (MAFF, 1979; Newbold, 1977; Robson and Barrett, 1977) (Chapter 6). Biological control is now being attempted with encouraging results (Chapter 5). For example, the Chinese Grass Carp (*Ctenopharyngodon idella*), a fast growing vegetarian fish, has been effective in reducing submerged weeds, providing human food in the process. However, the fish does not control emerged or free-floating weeds and because it cannot spawn outside captivity, repeated reintroduction is necessary (Anon, 1976). Another animal being considered is the Manatee or Sea Cow, a docile warm-blooded herbivore which could well be promising in tropical countries. Unfortunately, the Manatee is nearing extinction in the wild and has so far not bred in captivity. Crayfish, ducks and geese all have a useful place in aquatic weed control but none is likely to solve the large scale problems of free-floating weeds.

In India and Indonesia Water Hyacinth is harvested for direct feeding to cattle and pigs but the water content is too high to justify the cost of transport; methods of de-watering are therefore being developed so that advantage can be taken of the exceptional growth rates of aquatic weeds for cattle feed and other purposes (Anon, 1976).

There is now interest in North America in using Water Hyacinth to remove toxic and other mineral matter from polluted waters, or for processing raw sewage, using the harvested biomass to generate methane (Wolverton and McDonald, 1976).

## 4.2 POISONOUS PLANTS

Man has never been free from danger of poisons in his food and although in the developed world the dangers are now much reduced, his livestock continue to be at risk. In less fortunate areas poisonous plants and food containing mycotoxins still take a toll of human and animal life and cause considerable suffering.

Only occasionally are poisonous plants important in temperate horticulture, although some ornamental plants grown in parks and gardens are toxic. All parts of the Yew (*Taxus baccata*), for example, are poisonous, containing an alkaloid taxine which induces sudden death by stopping the pumping action of the heart; dead animals are sometimes found still to have Yew foliage in their mouths. Because animals eat the leaves so readily, Yew trees when they were grown for making long-bows were segregated from grazing animals in churchyards, where they are still found to this day. Generally, however, unless they are hungry or stressed, livestock tend to avoid most kinds of poisonous plant in the fresh condition; but they are unable to differentiate between safe and unsafe food when it is preserved in hay or silage. Poisonous plants that have been sprayed with herbicides often become more palatable while retaining toxicity. Livestock poisoning is relatively rare in Northern Europe, but in North America, Australia and in parts of Africa the total economic loss from poisonous plants is immense and kills of more than one hundred sheep from Halogeton (*Halogeton glomeratus*) alone are not uncommon in Utah, Nevada and Idaho.

Poisonous plants also cause injury, illness and pain. Unborn young may be aborted or born grotesquely deformed, unpigmented skin made acutely sensitive to sunlight, and in New Zealand awned seeds of Wall Barley (*Hordeum murinum*) work their way into the bodies of sheep causing severe pain. Repeated ingestion of poisonous weeds at sublethal levels can cause an unspecific malaise and unthriftiness which often goes undetected. Other plants may contain substances that are not harmful to the animal that eats them but meat or milk may be tainted or cause illness when consumed (Anon, 1968; Forsyth, 1968; Keeler *et al.*, 1978; Taylor, 1980).

### 4.2.1 Types of toxic action

Several plants produce non-poisonous glycosides that can be hydrolysed by enzymes within either the plant or the animal gut to release hydrocyanic acid (or prussic acid), one of the most rapid poisons known. Cherry Laurel (*Prunus laurocerasus*) can be fatal when eaten by livestock, but in Britain most poisoning cases with glycosides follow exessive consumption of crops such as kale and white clover. At sublethal doses these glycosides induce abortion or cause goitre. Man is less sensitive to prussic acid than are some of his animals, especially when sugar is consumed at the same time, and this has been suggested as the reason for the failure of the attempt on Rasputin's life in 1916.

Alkaloids are substances found in many plants which, although without any obvious function in the plant, produce physiological effects in animals. Some alkaloids are used in food, medicine and as pesticides, and plants containing them, including tea, coffee, poppies (for morphine), tobacco and Foxglove (for digitalin) are grown as crops. Hemlock (*Conium maculatum*) contains five different poisonous alkaloids, but principally coniine, all capable of inducing paralysis and death. Extracts of Hemlock have been used since ancient times in medicine and for homicides and Hemlock in the garden has occasionally been mistaken for parsley and consumed, with fatal results. In North America the Locoweeds (*Astragalus* spp.), the Western False Hellebore and Lupine (*Lupinus sericeus*) can induce cleft palates, median eyes and deformed legs in the unborn young of cattle when consumed in sublethal quantities.

The foliage of a number of crops such as rhubarb, spinach, the cultivated beets and several common weeds contain harmful levels of oxalic acid and soluble oxalates. When absorbed from the gut oxalic acid or its sodium or potassium salt neutralises blood calcium, which is harmful in itself, but then the needle-like calcium oxalate crystals block the bladder and urethra, especially in male animals; larger doses cause rapid death. In North America the most notorious oxalate plant is Halogeton (*H. glomeratus*) which accumulates oxalate through the growing season until in winter oxalate can contribute over 20 per cent of its dry weight; at this stage less than 300 g of foliage is fatal to a hungry sheep, and the weed is responsible for numerous 'sheep deaths' when flocks are being moved during winter. First noticed during the 1930s, this alien species from Russia has now spread throughout several Western states (James, 1978).

Another group of plants contain substances that induce extreme sensitivity to light, causing unpigmented skin to swell, crack and slough off. The worst single species is Perforate St. John's-wort or Klamath Weed (*Hypericum perforatum*), an innocuous wild plant in its native Europe which was introduced to North America, Australia and New Zealand in the last century, and more recently to South Africa. In all these countries it has spread aggressively in cattle range land. Wherever there is sufficient alternative food, cattle avoid eating St. John's-wort, but they are unable to do so when the weed is in hay or silage, and both active principles, the pigments hypericin and hypericum red, retain their activity.

Weeds contribute to the windborne pollen in the atmosphere which afflicts sufferers from hay fever. In Europe the wind-pollinated trees and the grasses are the main culprits, but in North America several kinds of Ragweeds (*Ambrosia* spp., mainly *A. artemesiifolia*) which are common roadside plants and occur in derelict urban areas, are equally important pollen sources, producing an estimated one million tonnes of pollen annually (King, 1966) (Chapter 1).

In addition to weeds, crops fed to inappropriate animals or at an unsuitable stage of growth can also kill and maim them. When grown in soils high in available nitrate, oat straw in small quantities can be lethal to all livestock, and both wheat and barley grain can cause death or lameness in horses. Potato tubers that have become green after exposure to light contain alkaloids poisonous to man and animals; tomato plants contain similar alkaloids and have been the cause of death of pigs.

A reasonable balance is required between ruthless destruction of all potentially hazardous plants and needless exposure of people and animals to danger. In Britain there is an extremely toxic plant, the Foxglove (*Digitalis purpurea*), present in most parts of the country, yet human or animal deaths from this source are rare. This is because the Foxglove, like many poisonous plants, tastes unpleasant and animals will not persevere with eating it. However, animals that have recovered from poisoning by a particular plant species may acquire the taste, or sometimes a desperate craving for that plant (Forsyth, 1968). Certain species, notably the Yew in Britain, are not unattractive to cattle and many other species which are shunned under normal circumstances may be eaten during drought, after herbicide spraying, when the grass is covered with snow or when animals are under extreme stress.

Plantings of ornamental species that produce attractive but poisonous seeds and fruits, such as Laburnum (*Laburnum anagyroides*) and Mezereon (*Daphne mezereum*), or whose foliage or sap can cause severe skin damage, such as Giant Hogweed (*Heracleum mantegazzianum*) or *Euphorbia* spp., should be avoided in domestic gardens and areas to which the public have free access. However, it is quite impossible to remove all dangerous plants and reliance must be placed on better education, especially of young children, to reduce the deaths and injury that still occur.

## 4.3 PARASITIC WEEDS

Parasitic flowering plants use crops as hosts, depriving them of water, mineral salts and in some cases of assimilates. There is also evidence that inhibitory substances are passed from the parasite to the host. Parasites may attach to the roots or to the stems, be completely dependent on the host or only require water and minerals from it. The flowering plant parasites are largely confined to the families and genera shown in *Table 4.2*.

### 4.3.1 Convolvulaceae

The Dodders (*Cuscuta* spp.) are stem parasites mainly attacking clovers, lucerne and other herbage legumes and commonly found wild on Gorse (*Ulex europaeus*) but found also on sugar beet, peppers (*Capsicum annuum*) and potato. Some species are restricted to a small host range, but most are catholic in their tastes, and in India one kind (*Cuscuta hyalina*) was found on 35 genera, while in Russia *C. lupulifor-*

*Table 4.2*

*Families and genera of parasitic plants*

| | *Family* | *Principal genera* |
|---|---|---|
| Stem parasites | Convolvulaceae | *Cuscuta* |
| | Lauraceae | *Cassytha* |
| | Loranthaceae | *Viscum* and *Phoradendron* |
| | | *Arceuthobium* |
| | | *Loranthus* |
| Root parasities | Orobanchaceae | *Orobanche* |
| | Scrophulariaceae | *Melasma* |
| | | *Rhinanthus* |
| | | *Striga* |

*mis* was found to parasitise 131 different crops and wild plants. Seedling Dodders survive for about a week without a host, but once a suitable one is found an attachment is made, growth and development of the weed is rapid, and seeds are produced after two or three months. The parasite smothers the host with its thin, intertwined, much branched stems which in a well-grown plant can reach a total length of over one kilometre (King, 1966). Control has been achieved in Britain by seed cleaning and by enacting regulations to prevent the sale of heavily contaminated crop seeds. Dodders are still considered potentially noxious weeds in Northern Europe and North America. In Eastern Europe and the Middle East yields of clover and sugar beet are reduced by Dodder infestation and herbicides are used regularly for its control (Anon, 1973).

### 4.3.2 Loranthaceae

This family contains the evergreen semiparasitic Mistletoes (*Viscum* spp. and *Phoradendron* spp.) that grow on the branches of many broad-leaved and some coniferous trees and the Dwarf Mistletoes (*Arceuthobium* spp.) that are mainly restricted to conifers but can occur on walnuts, as they do in California. The family also contains the highly destructive tropical species still generally classified under *Loranthus* but whose nomenclature is under review (Weeraratna, 1960).

The deleterious effects of the semiparasitic true Mistletoes are largely confined to withdrawal of water and shading, but this can be severely debilitating in Southern Europe, especially on pears and some plantation crops (Preston, 1977). Hosts invariably suffer severely, however, when parasitised by the Dwarf Mistletoes (Leonard, 1973) because not only is water extracted but assimilates are diverted from a host to a parasite sink. The effect is to reduce growth and deform the trees, sometimes actually killing them. Losses in timber production due to these parasites in North and Central America have been estimated at 20 million cubic metres annually. Dwarf Mistletoe species are most numerous in the New World, but there are four species in Africa, Europe and Asia (Hawksworth, 1973). The individual parasitic plant is always small, and in one species (*A. minutissimum*) is only 5 mm long when fully grown, making this the smallest dicotyledonous species (King, 1966). Despite the plant's small size, the sticky seeds are ejected from the fruit with such force that they can travel as far as 13 m horizontally which enables the seeds to reach needles and twigs of a new host. After host penetration by the radicle the cycle begins once more.

The genus *Loranthus* contains numerous kinds of stem parasites that cause severe damage to tropical timber trees (including teak), to citrus, rubber and other plantation crops, and to tea bushes. In tea the leaves of the parasite are superficially like those of the crop and are sometimes picked by mistake, giving the tea an unpleasant bitter taste. These parasites are spread by birds to new plantations from old or abandoned ones, and one method suggested for their control is to leave a strip of forest around new plantings to attract birds (King, 1966), in much the same way that fruitgrowers in Britain sometimes retain an area of old and unproductive cherries to draw birds away from other more valuable fruit crops.

### 4.3.3 Orobanchaceae and Scrophulariaceae

The parasitic members of Orobanchaceae are root parasites which after attachment to the host's roots emerge above ground and after flowering liberate numerous small seeds. These are easily dispersed by wind or in irrigation water, and Broomrape (*Orobanche*) seed is also dispersed by cattle and goats that eat the flower spikes. The economically significant species occur in hot climates in Asia, Africa, Southern Europe and North America, and apart from two genera (*Aeginetia* and *Christisonia*) with species parasitising sugar cane in the Far East they are restricted to species of *Orobanche* and *Striga*.

Of the hundred or so species of Broomrape (*Orobanche*) many parasitise wild species and are of little economic importance; in Britain, for instance, there are eleven native species but none occurs on crops. The most troublesome species found as weeds on crops are *O. crenata* on broad beans, sometimes also on carrots, tomatoes and even on Bermuda Buttercup (*Oxalis pes-caprae*) which is a widespread weed in countries bordering the Mediterranean; *O. cumana* (= *O. cernua*) on sunflowers in *Eastern Europe; O. ramosa* on tobacco and tomatoes in Europe, the Middle East, North Africa and more recently on tomatoes following its introduction to California; and O.

*aegyptiaca* which prefers the higher temperatures of Northern Africa and has a similar host range to *O. cumana*.

It has been shown that the Broomrape (*O. crenata*), on broad bean deprives the host of water by maintaining a high osmotic potential in its tissues, and that this lack of water in the host is responsible for the reduction in vigour and yield (Whitney, 1973). Application of 2,4-D to the host foliage is slowly moved down to the parasite, where it accumulates preferentially. However, water moves very rapidly from the soil to the parasite via the host root system which could provide fast and effective weed control with a soil-acting herbicide.

The small seeds of the Broomrapes easily contaminate crop seed, especially when that is also small, as it is in tobacco, making separation virtually impossible. Some control of Broomrapes is achieved by using bait crops in rotation with susceptible ones but considerable losses still occur.

The Witchweeds (*Striga* spp.) (Holm *et al.*, 1977) differ from the Broomrapes (*Orobanche* spp.) in having some green leaves, although prior to emergence they derive all their assimilates from the host. Another major difference lies in the host range, for Broomrapes never parasitise grasses, yet grasses, including maize, sorghums and sugar cane, are the crops most often attacked by Witchweeds. Research into Witchweeds was intensified following the discovery in 1956 of *Striga asiatica* (= *S. lutea*) in North and South Carolina. The potential threat to grain production was quickly recognised and efforts were made to contain the outbreak, but the weed had already been spread to over one hundred locations in the Carolinas. Quarantines and controls were imposed within the infested area. Herbicides have been regularly and intensively used to control Witchweeds and weed grasses, and trap crops that stimulate germination without themselves becoming infected have been grown in preference to susceptible ones; ethylene has also been injected into the soil in a further attempt to stimulate dormant seeds to germinate and then die in the absence of a host. The spread of Witchweeds in North America has been halted, the levels of infection in crops have been reduced and eventual eradication is now confidently anticipated.

In other tropical and subtropical areas of the world such an intensive and expensive control programme is impossible, and means are sought whereby reasonable crops can be obtained despite competition from Witchweeds. Ogborn (1972) devised a cheap and effective way of controlling the above ground parts of *S. hermonthica* in grain sorghum by spot spraying herbicide through a simple hand-held plastic sprayer, which resulted in substantial yield increases compared with the normal method of hand pulling. But to be really effective a chemical treatment should prevent the parasite from attaching to the host, because much of the crop damage occurs before emergence while the Witchweed is extracting sufficient assimilate from the host to support flower and seed production (Kasasian and Parker, 1971).

The active ingredient in the host root exudate causing Witchweed germination, strigol, has been isolated and chemically characterised. Synthetic analogues have been made which are even more active than strigol, and stimulate germination of dormant seeds at the low concentration of $10^{-9}$M. Experiments with these compounds in India have shown that a single incorporation in the soil at between 0.75 kg/ha and 0.075 kg/ha can halve the remaining numbers of Witchweed seeds (Johnson *et al.*, 1976). Similar results have been obtained with these compounds on dormant seeds of *Orobanche ramosa*. However, for economic and other reasons this technique is not yet practicable.

## 4.4 CROPS AS WEEDS

In Britain volunteer plants arising from seeds or tubers of potatoes, wheat, barley and oil-seed rape frequently become weeds in subsequent crops. Potentially more serious are problems caused by hybrids or subspecies of crops, that are useless in themselves but are superficially indistinguishable from the crop. Already considered in an earlier chapter, is another category, the useful crop plants which, when introduced to a new environment, grow and reproduce so prolifically that they become weeds.

Volunteer wheat and barley plants which germinate before either autumn or spring sown cereals become infected by pathogens causing leaf diseases thereby acting as a bridge to pass on infection from the old crop to the new. The economic importance of mildews and rusts in cereals has increased rapidly and it has been claimed that cereal volunteer plants are among the most costly of all weeds in Britain. Further harm is done when seed from volunteer plants is harvested, for not only does it reduce purity of seed crops, but when they are found to occur at above 3 per cent, which at present is not uncommon, the grain sample will fail to meet new EEC requirements under Intervention Schemes.

Whenever potato crops are harvested, many tubers remain in the soil; late cultivars can also produce viable seeds in favourable seasons but potato seedlings seldom pose a difficult problem. Plants growing in subsequent years from tubers compete strongly with slow growing row crops such as carrots, beet, onions and brassicas, and in peas grown for processing parts of potato plants can cause potentially dangerous contamination. Volunteer potatoes, known as groundkeepers, are especially important in the perpetuation of leaf-roll and other viruses, blackleg, gangrene, potato cyst eelworm and in North America and Europe but not in Britain, of Colorado beetle.

The size of the potato groundkeeper problem can be immense, a survey in East Anglia showing that in 1973 between 120,000 and 370,000 tubers/ha, most of which were under 4 cm in diameter, remained in the soil after mechanical harvesting (Lutman, 1977); in The Netherlands up to 460,000/ha have been found. In the British Survey about one third of the tubers were on the surface, another third were in the upper 5 cm while the remainder were buried more deeply. In three relatively mild winters, but during which the surface temperature of the soil fell to –8°C, the temperature at 5 cm never fell to freezing point, and over half of the buried tubers survived the winter. Groundkeeper potatoes cause problems in the next crop and, unless controlled, in subsequent ones too, because each plant can produce up to three small new tubers despite competition from a crop. Chemical control by applying herbicides to the foliage is not fully successful unless repeated applications can be made, because the shoots emerge continuously from April to June according to the depth of burial (Lutman and Richardson, 1978).

Plants that combine the properties of noxious weeds with the visual appearance of a crop pose peculiar problems in control. In an earlier chapter the problem of annual beet was described, and in North America crop sorghums have sometimes been accidentally pollinated by Johnson-grass (*Sorghum halepense*); in both cases commercial seed that had become contaminated during pollination, and was therefore hybrid, initiated the problem. In another example from North America, the Dudaim melon, which was probably introduced as a potential donor of useful genes for pest and disease resistance, has itself become a weed.

A small annual melon, the Dudaim (*Cucumis melo* var. *dudaim*), with fruits known as pepos a mere 8 cm long, was taken from Africa to India where it became known as Queen Anne's Pocket Melon and is now grown both as an ornamental and for food. Dudaim melons reached Imperial Valley in California from India in 1953, and by 1955 escaped plants were common in some parts of California and the plant had also been taken to Mexico by returning itinerant workers who had acquired a taste for it. Many people who eat the fruit suffer from gastritis but not, apparently, Mexicans.

The potential of this vigorous plant as a weed was soon recognised but despite determined efforts at containment it continued to spread. Detached pepos containing seed can float and become dispersed in irrigation and drainage channels, and if they remain whole may stay in the soil for many years without loss of viability of the seeds inside. Most infestations have been in asparagus, which effectively hides the weed until it has begun to produce fruits, but it also grows well in other crops. Each plant can produce several vines up to 8 m long and can ripen upwards of 100 pepos. In Mexico infestation is in grazing land where dissemination of seed is accomplished by cattle which eat the fruits and excrete viable seeds.

A more insidious problem is created where Cantaloupe melon seed is produced in areas infested with the Dudaim. The two plants hybridise freely, producing in the $F_1$ and $F_2$ generations seeds and seedlings superficially indistinguishable from normal ones. The fruits on hybrid plants are sweet smelling but are bitter to the taste (Dixon and Kreps, 1973).

A possible future problem is the establishment as a weed of a crop that is inherently resistant to soil-acting herbicides. When persistent herbicides are applied repeatedly in a perennial crop, the crop seedlings often tend to survive. Thus blackcurrant seedlings are common in that crop, and asparagus grows freely from seeds shed in asparagus beds. So far, however, no perennial crop seedlings have created more than local and temporary problems.

## 4.5 WOODY WEEDS

Trees provide shade and shelter for foraging animals, especially during drought, and so long as they do not encroach on grazing land are beneficial. But trees and shrubs often respond to overgrazing and to bush fires by extending into the grassland; cutting back then seems the only practical way of preventing the grass being taken over. In Africa young *Acacia* trees respond to cutting by springing up from the base and competing even more strongly with the grass. Some of the *Acacias* are vigorous, thorn covered scrambling climbers which become particularly prevalent when their natural competitors are destroyed during bush clearance for pasture. Goats can provide a means of controlling woody weeds but they must be prevented from going on to destroy the sward also (Ivens, 1960).

In New Zealand much of the pasture land has been obtained by clearance of climax rainforest and so it is perhaps not surprising that many of the weeds are woody, some native, but most introduced. The most spectacular is Gorse (*Ulex europaeus*), which grows to over 6 m tall with a luxuriance and density unknown in its native Europe; it was originally introduced for hedging and livestock could graze on the young growth. Another is Sweet Briar (*Rosa rubiginosa* = *R. eglantaria*) introduced from England as an attractive garden plant, but which now forms dense thorny clumps in pasture land, especially since the recent decline in rabbit numbers.

Another group of woody weeds in New Zealand are the Willows (*Salix*. spp.), especially the Crack Willow (*S. fragilis*), which have spread rapidly and become a severe problem largely during this century. Large areas of low lying pasture land have been converted into useless swamps by Willows blocking the streams with their prolific growth.

Small pieces of the Crack Willow easily become detached, float downstream and begin a new infestation (Little, 1960).

Threatophytes are plants that grow in arid areas where they waste precious water by rooting into ground water, often in valley bottoms beside streams or bordering irrigation canals. For example, in the Western United States, Willows (*Salix* spp.), Poplars (*Populus* spp.) (known in North America as Cottonwoods), Mesquite (*Prosopis juliflora*) and the Salt-cedars (*Tamarix* spp.) are all woody weeds that cause excessive loss of water in areas where water is scarce and valuable.

## REFERENCES


ANON, (1968). *22 plants poisonous to livestock in the Western States*, United States Department of Agriculture, Agriculture Information Bulletin 327, Washington D.C., p. 64

ANON, (1971). *Economic damage caused by aquatic weeds* (preliminary survey), Office of Science and Technology, Agency for International Development, Washington, p. 13

ANON, (1973). *Proceedings of the European Weed Research Council Symposium on Parasitic Weeds, 1973*, EWRC, Wageningen, p. 295

ANON, (1976). *Making aquatic weeds useful – some perspectives for developing countries*, National Academy of Science, Washington D.C., p. 175

DIXON, D. and KREPS, L. (1973). Dudaim melon – a direct threat to asparagus, *Proceedings 25th Annual California Weed Control Conference*, pp. 37–42

FORSYTH, A. A. (1968). *British poisonous plants*, MAFF bulletin 161, HMSO, London, p. 131

FRYER, J. D. and MAKEPEACE, R. J. (eds.) (1977). *Weed Control Handbook*, Vol. I, *Principles*, Blackwell, Oxford, p. 510

GAUDET, J. J. (1979). Aquatic weeds in African man-made lakes, *PANS*, **25**, 279–286

HAWKINS, A. F. (1972). *Control of algae, Outl. Agric.*, **7**, 21–26.

HAWKSWORTH, F. G. (1973). Dwarf mistletoes (*Arceuthobium*) of coniferous forests of the world, in *Proceedings European Weed Research Council Symposium on Parasitic Weeds, 1973*, EWRC, Wageningen, p. 295

HOLM, L. (1977). Weeds and water in world food production, *Weed Sci.*, **25**, 338–342

HOLM, L. and YEO, R. (1979). The biology and control of aquatic weeds, in *Les Desherbage des Cultures sons les Tropiques* (eds. J. Deuse and E. M. Lavabre), Maisonneuve et Larose, Paris, p. 312

IVENS, G. W. (1960). Species of Acacias as weeds, in *The Biology of Weeds* (ed. J. L. Harper), Blackwell, Oxford, p. 256

JAMES, Lynn F. (1978). Oxalate poisoning in livestock, in *Effects of Poisonous Plants on Livestock* (eds. R. F. Keeler, K. R. Van Kampen and Lynn F. James), Academic Press, New York, p. 600

JOHNSON, A. W., ROSEBERRY, G. and PARKER, C. (1976). A novel approach to *Striga* and *Orobanche* control using synthetic stimulants, *Weed Res.*, **16**, 223–227

KASASIAN, L. and PARKER, C. (1971). The effects of numerous herbicides on the germination of *Orobanche aegyptica* and *Striga hermontheca, PANS*, **17**, 471–481

KEELER, R. F., VAN KAMPEN, K. R. and JAMES, L. F. (eds.) (1978). *Effects of Poisonous Plants on Livestock*, Academic Press, New York, p. 600

KELLEY, A. and BRUNS, V. (1975). Dissemination of weed seeds by irrigation water, *Weed Sci*, **23**, 486–493

LEONARD, O. A. (1973). Translocation in and between mistletoes and their hosts and the significance of this in relation to weed control, in *Symposium on Parasite Weeds*, Malta, 1973, EWRC, Wageningen, 188–193

LITTLE, E. C. S. (1960). The ecology of some New Zealand woody weeds, in *The Biology of weeds* (ed. J. L. Harper), Blackwell, Oxford, p. 256

LUTMAN, P. J. W. (1977). Investigations into some aspects of the biology of potatoes as weeds, *Weed Res.*, **17**, 123–132

LUTMAN, P. J. W. and RICHARDSON, W. G. (1978). The activity of glyphosate and aminotriazole against volunteer potato plants and their daughter tubers, *Weed Res.*, **18**, 65–70

MAFF, (1979). *Guidelines For the Use of Herbicides on Weeds in or Near Watercourses and Lakes*, Ministry of Agriculture, Fisheries and Food, booklet 2078, London, p. 55

MITCHELL, D. S. (1972). The Kariba Weed: *Salvinia molesta, Brit. Fern Gaz.*, **10**, 251–252

NEWBOLD, C. (1977). Aquatic herbicides: possible future developments, in *Ecological Effects of Pesticides* (ed. F. H. Perring and K. Mellanby) Academic Press, London, p. 192

OGBORN, J. E. A. (1972). The control of *Striga hermonthica* in peasant farming, *Proceedings 11th British Weed Control Conference*, pp. 1068–1077

PRESTON, A. P. (1977). Effects of mistletoe (*Viscum album*) on young apple trees, *Hort. Res.*, **17**, 33–38

ROBSON, T. O. and BARRETT, P. R. F. (1977). Review of effects of aquatic herbicides, in *Ecological Effects of Pesticides* (ed. F. H. Perring and K. Mellanby), Academic Press, London, p. 193

TAYLOR, J. A. (1980). Bracken: an increasing problem and a threat to health, *Outl. Agric.*, **10**, 298–304

THOMAS, K. J. (1979). The extent of *Salvinia* infestation in Kerala (S. India): its impact and suggested methods of control, *Environmental Conservation*, **6**, 63–69

TIMMER, C. E. and WELDON, L. W. (1967). Evapotranspiration and pollution of water by Water Hyacinth, *Hyacinth Control Journal*, **6**, 34–37

WEERARATNA, W. G. (1960). The ecology and biology of parasitism of the Loranthaceae of Ceylon, in *The Biology of Weeds* (ed. J. L. Harper), Blackwells, Oxford, p. 256

WHITNEY, P. J. (1973). Transport across the region of fusion between bean (*Vicia faba*) and broomrape (*Orobanche crenata*), in *Proceedings European Weed Research Council Symposium on Parasitic Weeds, 1973*, EWRC, Wageningen, p. 295

WOLVERTON, W. and McDONALD, R. C. (1976). Don't waste waterweeds, *New Scientist*, **71** (1013), 318–320

## FURTHER READING

HOLM, L., PLUCKNETT, D. L., PANCHO, J. V. and HERBERGER, J. P. (1977). *The World's Worst Weeds*, University Press of Hawaii, Honolulu, p. 609

KASASIAN, L. (1971). *Weed Control in the Tropics*, Leonard Hill, London, p. 307

KING, L. J. (1966). *Weeds of the World: Biology and Control*, Leonard Hill, London, p. 526

KINGHORN, A. D. (ed.) (1979). *Toxic Plants*, Columbia University Press, New York, p. 195

MUSSELMAN, L. J. (1980). The Biology of *Striga*, *Orobanche*, and other root-parasite weeds, *Ann. Rev. Phytopathol.*, **18**, 463–489

RUSSELL, G. E. (1978). Resistance to parasitic weeds, in *Plant Breeding for Pest and Disease Resistance* (ed. G. E. Russell), Butterworths, London, p. 485

SCULTHORPE, C. D. (1967). *The Biology of Aquatic Vascular Plants*, Edward Arnold, London, p. 610

# 5 WEED CONTROL WITHOUT HERBICIDES

## 5.1 INTEGRATED CONTROL

Following the discovery of potent insecticides and herbicides in the 1940s a euphoric attitude prevailed in which chemicals were seen as the panacea of all ills, able to provide easy and effective control of pest problems. Soon, however, insect resistance required the application of higher rates, destruction of predators increased the severity of pest infestations and sometimes actually created new pests. A reappraisal of chemotherapy was begun as the ecological implications for wild life and man himself of the highly persistent pesticides then in use became known. Gradually the notion of integrated control has evolved, in which pest and disease infestations are reduced below the economic threshold by a judicious combination of all feasible means, chemical, biological or cultural. The life cycles of pests and their parasites and predators were re-examined and improved pesticide application methods and pest and disease forecasting techniques were devised to provide an adequate control cheaply and with the least possible contamination of non-target organisms.

Weed control, however it is achieved, should be treated as an integral part of good husbandry, but only recently has the concept of integrated control been applied with conviction to the control of weeds, although the problems are essentially similar to those met in dealing with pests and pathogens (Fryer and Matsunaka, 1977); these include herbicide resistant floras of grasses, sedges and other problem weeds which have arisen by interspecific selection and an increasing dependence on herbicides for weed control in most developed countries. Furthermore, the occurrence of interaction between presence of insecticide residues and herbicide persistence times (Chapter 6) and between weeds, pests and disease organisms (Chapter 1), and of more direct effects of herbicides on the incidence of disease (Altman and Campbell, 1977) make an integrated approach to these problems essential.

In spite of intensive hand-weeding, control of weeds in the past was probably much less effective than it is today. For example, there are many historical references to the toxic effects of Darnel and

Corncockle contamination of wheat in Northern Europe, yet both weeds have now almost disappeared. This improvement began with innovations such as efficient mechanical separation of crop from weed seeds, more uniform germination resulting from better drills, protection from pests and diseases, seed of guaranteed high germinability and use of fertilisers to increase early crop growth; thus herbicides are just one more of the factors in this general improvement. Plant breeding has produced more vigorous crops, but some of the recent cultivars, such as the short strawed cereals, are dependent on chemical weed control for their success. However, the alternative methods of controlling weeds, such as cover-cropping and mulching, crop rotation, flooding, burning and intensive use of pigs and other animals can often be combined with herbicides to obtain a better result more economically. In addition, research in many countries is attempting the control of certain problem weeds biologically, using insects, fish, viruses and parasitic fungi.

## 5.2 CULTIVATION

Cultivation for weed control, most of it by hand, has been claimed to be the most time consuming and exhausting human occupation; certainly, it remains the most important method of weed control everywhere except in the developed nations. According to Fryer and Makepeace (1977), cultivations control weeds by burying them, by cutting them off near soil level, by stimulating weed seeds to germinate so that they can be destroyed, by desiccation of roots and rhizomes brought to the surface and finally by continually cutting back the growth of deep rooted perennials, thereby exhausting them. It is now believed that soil disturbance beyond the minimum necessary for weed control often reduces yields and that provided weed infestation can be prevented in some other way much of the traditional soil disturbance can be dispensed with (Chapter 8). However, well-timed overall cultivations can be a cheap and effective method of killing weed seedlings that can be combined with residual herbicide treatment for prolonged weed control. In the stale seedbed technique a tilth is prepared in the usual way, but by delaying drilling of a crop for two or three weeks the initial flush of weed seedlings is easily destroyed with a contact herbicide before the crop germinates. The roots and rhizomes of many perennial weeds cannot withstand prolonged desiccation, and infestations of both Common Couch and the tropical Purple and Yellow Nutsedges (*Cyperus rotundus* and *C. esculentus*) can be reduced by exposure to sun and wind following cultivation.

Ploughing has been the primary process in the creation of a seedbed for over 2,000 years. Before the development of modern iron and steel ploughs, the depth achieved was shallow and erratic and inversion of the slice was incomplete, so that burial of weeds was often poor. Harrows and tined cultivators have been in use for an equally long time for the breaking down into a seedbed of a rough ploughed surface. Recent developments include powered cultivators, some of which produce a seedbed tilth without ploughing. Fast tractor cultivating enables the soil to be worked when in the optimum physical condition, and the cutting and shattering of weed roots and rhizomes is more complete.

In peasant agriculture, the hand hoe remains responsible for most weed control, the area that a local community can keep clean during the early and critical stages of crop growth often determining how much food it can produce.

## 5.3 FLOODING

Paddy rice is able to thrive partially submerged under water and weed control is achieved by a combination of careful management of water depth, cultivations and, in Japan and some other developed countries, application of herbicides. One of the most competitive and universal of all rice weeds is the annual Barnyard-grass (*Echinochloa crus-galli*) which is capable of reducing yields at modest levels of infestation by almost 50 per cent. Fortunately the germination percentage declines from almost 100 to less than one when there is more than 20 cm depth of water, although a variety of this grass able to germinate under water has evolved in some areas. Occasional drainage of the paddy allows algae to become desiccated and die (Kasasian, 1971; Noda, 1977).

Flooding, however, is rarely a practicable method of weed control in crops other than rice.

## 5.4 BURNING

Fire has been the main tool in scrub clearance before land is taken into cultivation for many thousands of years and it remains a cheap and effective method. However, in tropical forests on soils poor in bases, most of the mineral nutrients are contained in the biomass. Burning is soon followed by leaching from the ash which impoverishes the soil so that it is unable to sustain sufficient ground cover to prevent erosion. Fire is effective also in disposing of unwanted organic debris in crops such as strawberries, cereals and sugar cane. Strawberry leaves do not form an abscission layer and so cannot shed their leaves, and the old foliage which accumulates at the end of the harvest is sometimes destroyed by firing both the dead leaves and straw; firing may sometimes also stimulate more fruit production in the next year. Although the rapid burn is spectacular, the temperatures reached are only locally sufficient to destroy weed seeds and the fire does little to aid weed control in this crop. In continuous cereal growing, burning is the simplest and most effective way of disposing of the straw residues and stubble which interfere directly with drilling (especially direct drilling) and indirectly through anaerobic decomposition products that are toxic to germinating crop seeds. It might be expected that weed seeds in the stubble would be destroyed by burning but studies with weed grasses suggest that, as with strawberries, many weed seeds survive unharmed (Cussans and Wilson, 1976). Sugar cane is sometimes burnt over prior to cutting to destroy snakes and vermin and to remove the 'trash' of old leaves, but because inversion of the sucrose begins rapidly, the field once burnt must soon be cut. Fire is a useful but dangerous and unpredictable tool and wind direction changes can cause it to leapfrog fire breaks and spread to adjoining urban and forest areas. Despite its economic advantages stubble burning in a crowded community like Britain probably cannot be tolerated during droughts when the general risk of fire is high.

### 5.4.1 Selective weed control by flaming

Most plants cannot survive prolonged exposure above 40°C, although thermophiles from hot deserts and tropical forests have become adapted to higher temperatures. Blue–green algae from hot springs, for instance, can actually grow above 90°C. A few higher plants can survive at above 55°C and, exceptionally, some Cacti (*Opuntia* spp.) still grow at 65°C (Sutcliffe, 1977). When the period of exposure is short, correspondingly higher temperatures are necessary to kill plants, and seeds are considerably more resistant than growing plants (Crafts, 1975).

LPG (liquefied petroleum gas) or oil burners can be used for weed control by flaming in crops. This is often done in two operations, some days apart, the first sufficient to cause wilting and the second to destroy the dying plants. Selectivity is possible in onion and some others crops but the margin of safety is small and flaming is only practicable where oil or gas are cheap; for most situations herbicides are cheaper and safer, but burning does have the advantage of leaving a negligible residue.

For domestic gardens, small hand-held flame guns are advocated but they are expensive and potentially dangerous and herbicides are probably superior in most respects.

## 5.5 SOIL STERILISATION

Soil can be partially sterilised chemically with methyl bromide or dazomet, or with steam, to control soilborne pests and pathogens; the control of weeds and weed seeds is usually a bonus. Sterilisation is expensive and is used in Britain only on high value crops, particularly under glass, but large areas of horticultural monocropping in California are now chemically sterilised to reduce damage from soilborne pathogens. The increasing prevalence in Britain of intractable soilborne problems such as club root of brassica crops will encourage more soil sterilisation on a field scale. Specialised horticultural uses of soil sterilisation for weed control include the chemical pre-treatment of soil before sowing of high quality lawns and steam sterilisation of soil for container-grown nursery stock and pot plants.

## 5.6 MULCHING

Annual weeds can be partially controlled by organic mulches of straw, peat, bark chips, sawdust and other waste products or more recently with black polythene. Straw, for example, aids weed control under fruit trees in eastern England, although its primary purpose is water conservation. However, many types of perennial weed thrive under loose organic mulches, and although they are generally unable to penetrate black polythene, the tropical Purple Nutsedge or Nutgrass (*Cyperus rotundus*) can penetrate sheeting that is thinner than 1000 gauge (254$\mu$m) (Henson and Little, 1969). Polythene mulching is used in Britain for strawberries under plastic tunnels and for other small scale crops, but in Hawaii pineapples are planted through pre-laid sheeting which then controls almost all weeds in the covered area for the three years duration of the crop. As equipment for laying plastic film is simplified its use in high value crops will increase.

A rather different application of polythene sheet for weed control, now used in California and Israel, is to lay a continuous cover of white polythene during the hottest part of the year; at first perennial weeds grow but as the temperatures continue to rise they are killed by heat, and after two or three months the sheeting can be removed for re-use elsewhere.

A new type of mulch on trial in West Africa uses a living ground cover crop into which maize is planted. The cover is a scrambling perennial legume, such as Centro (*Centrosema pubescens*), and strips 15 cm wide are sprayed with paraquat before drilling. A growth retardant is then sprayed soon after emergence to check the growth of the cover crop and prevent it climbing over the maize. Results from such a 'living mulch' suggest that the protection from erosion and from organic matter loss, which are advantages associated with ordinary organic mulches, can be combined with good control of weeds without the need for cultivation. The ground cover crop also provides nitrogen fixation of up to 100 kg/ha in the root nodules.

In shifting cultivation it is loss of fertility and build up of weeds that forces the farmer to abandon his land and move on. It is suggested that the 'living mulch' could be effective in delaying this move, thus allowing more time in the recuperative phase of the shifting cultivation cycle (Akobundu, 1980) (Chapters 10 and 11).

## 5.7 BIOLOGICAL CONTROL

The need to reduce pesticide residues and the increase in resistance of insects to insecticides has encouraged research into biological control. Compared with work on insect pests a modest effort only has so far been put into the biological control of weeds. There have been some outstanding successes, and work is continuing on almost 80 weed species in North America alone (Andres, 1977*a*). Whenever a new organism is introduced for biological control there is always a nagging worry that it may adapt to a related crop or wild plant, then becoming a new pest. The most thorough investigations of the feeding habits and life history of the control agent are therefore essential before it is released (Anon, 1978).

### 5.7.1 Biological control with insects

The prime requirement of a biological control agent is that it should stress the weed population without having undesirable effects on useful plants. Many plant-eating insects are extremely selective in food preference and will starve rather than eat alternative food provided, but the great diversity of insects makes it probable that some species can be found that lives on the noxious weed it is wished to control. Alien weeds have generally been introduced without the insects that formerly controlled their populations, and the successful weed control programmes against Prickly Pear in Australia and St. John's-wort and Alligator-weed in North America have relied on restoring the natural insect controls to the introduced weed. The weeds that have attracted most interest as candidates for biological control are alien species, generally perennials, growing in dense patches on undisturbed low value land; in addition, there are many programmes directed against aquatic weeds.

Two species of Prickly Pear (*Opuntia inermis* and *O. stricta*) spread rapidly in Queensland and New South Wales (Australia) after introduction from South America, and at their peak in 1925 were estimated to cover nearly 30 million hectares of grazing land. Entomologists sent to South America and other countries in 1920 returned with 145 insects able to feed on *Opuntia* spp. Their feeding habits were revealed by using starvation tests, and animals that could exist on any other plants were rejected, leaving only 18 candidate insects. When tested in the field, one of these, the larva of a moth from the Argentine, *Cactoblastis cactorum*, seemed much the most effective. The moth lays about 75 eggs in a 'stick' attached to the cactus spines, which then hatch out and burrow into the cladodes, entirely destroying the soft inner tissues; the plants then collapse and become invaded by bacteria and fungi. In 1926 and 1927 several million eggs were collected from *Cactoblastis* moths reared in captivity and attached to Prickly Pear plants, and by 1929 large areas appeared to be dead and the insect was spreading rapidly; however, new growths of cactus from the base of 'killed' plants appeared, and it was not until 1935 that the invading cacti had been almost wholly destroyed (Newman, 1965).

As might be expected, it has often been possible to find insect enemies of introduced weeds, but there have been cases where an indigenous plant has succumbed to an introduced insect. For example, on Santa Cruz, a small island off the coast of California, native *Opuntia* spp. were encouraged by severe overgrazing to spread and take over much of the pasture. A substantial degree of control has now been achieved by introducing the Cochineal insect (*Dactylopius opuntiae*) from Hawaii. A rather similar situation prevailed in New Zealand when in about 1942 the native Manuka or Ti-tree (*Leptospermum scoparium*), a weed of upland pastures, became colonised by a previously unknown species of mealy bug, which had probably been blown by wind from Australia. However, by 1959 a fungus was attacking the mealy bug, so that the Manuka was once more able to thrive. This example illustrates the need for insect candidates for biological control to be themselves freed from natural enemies if their introduction is to push the equilibrium far enough towards controlling the weed population.

The nature of the damage done by insects in part determines the degree of control achieved. Defoliation by leaf feeders, although spectacular, may be much less permanently harmful to weeds than the destruction of vital and irreplaceable tissues by stem and root borers, such as the *Cactoblastis* moth larvae. Annual species, on the other hand, may be controlled most effectively by insects that destroy flowers and seeds. Although insects alone may not provide sufficient control the stress they apply to the weed population may be additive to the effects of other organisms or other control measures. For example, use of herbicides on mixed aquatic weeds in North America tended to favour the resistant and aggressive Alligator-weed (*Alternanthera phylloxeroides*) so that it quickly increased and occupied the whole stream; biological control of this weed achieved by a leaf and stem-eating Flea beetle (*Agasicles* sp.) collected in South America, which has been extremely successful, has made herbicide treatments more effective.

### 5.7.2 Fungal parasites for biological control

Other organisms that have been considered for biological control include fungal and bacterial parasites, snails and grass-eating fish. Among the fungi, those responsible for rust diseases often show the degree of host specificity that is essential if they are to be safely released into a new environment, but the possibility, however remote, of a pathogen producing a mutant form able to parasitise a crop must always be considered.

Some success has been claimed with the introduction of parasitic fungi against Bramble (*Rubus fruticosus*) in Natal, Dodder (*Cuscuta* spp.) in Russia and Water Hyacinth (*Eichhornia crassipes*) in India, and attempts have been made in New Zealand to increase the virulence and spread of an existing rust fungus present on Creeping Thistle (*Cirsium arvense*). However, the most thoroughly researched and successful campaign has been the attempt to control Skeleton-weed in Southeast Australia. This plant, which originated in the Mediterranean region and Southern Russia where it is relatively harmless, became a noxious weed in wheat cultivation on introduction to Australia. On the assumption that its changed status arose in part from a lack of biological control, a search for natural enemies was made in Europe. The most promising appeared to be *Puccinia chondrillina*, a monoecious rust that occurs throughout the plant's range in Europe, and can parasitise all above ground parts of the plant throughout the year (Hasan, 1974). The Australian form of Skeleton-weed, which probably arose from a single introduction and is genetically uniform, was found to be especially susceptible to a strain of the fungus found in Southeast Italy; observations in Europe also showed that in those areas with a climate similar to that in Southeast Australia, the rust was a dominant influence on Skeleton-weed populations. After testing the rust fungus in Europe to ensure that lettuce and other related crops were immune, uredospores of *P. chondrillina* were released in Australia. The fungus spread rapidly, extending over 300 km in 12 generations, and damage to seedlings and larger plants has been spectacular (Cullen, *et al.* 1973). This success has encouraged the study of similar specific rust diseases that have been recorded on other problem weeds such as St. John's-wort (*Hypericum perforatum*) and Lantana (*Lantana camara*).

### 5.7.3 Practical limitations of biological weed control

Biological control programmes have attempted to reduce the prevalence of single troublesome weed species, either invaders of large areas of low-value undisturbed land or aquatic weeds. The conditions of arable farming and horticulture, in which the space left by reduced numbers of any single weed species is taken up rapidly by others, do not lend themselves to biological weed control. However, experience with the *Puccinia* rust on Skeleton-weed, a weed of arable wheat fields, suggests that microorganisms may be suitable for more varied weed problems than the 'classical' method of introducing an insect from the natural range of the weed.

Instead of relying on the natural spread and increase to epidemic levels of plant pathogens, spores can be collected and applied in massive inoculations as a 'mycoherbicide' (Templeton, *et al.*, 1979). Both native and introduced pathogens can be considered and it is claimed that if the natural constraints on the fungus can be overcome then almost complete control of a particular noxious weed should be possible.

The major limitation on the use of introduced natural enemies, however, must be the protection of crops and socially desirable plants from an introduced organism that turned its attention away from weeds, and Wapshere (1975) has proposed a code of practice based on a phylogenetic method of testing. This should ensure that a biological control agent is entirely safe before it is introduced by progressively testing plants more and more distantly related to the problem weed. In the case of Skeleton-weed (*Chondrilla juncea*), the only crop grown in Australia that is in the same Chicoreaceae subfamily of Compositae is lettuce; therefore all candidate control agents for Skeleton-weed control showing any tendency to attack lettuce were discarded.

Weeds are plants in the wrong place and even the most noxious kinds may exhibit a dual personality, so that in other circumstances the vigour and rapid spread that make them feared as weeds become valuable assets. The Manuka (*Leptospermum scoparium*) in New Zealand, for example, is an aggressive invader of upland pastures, yet the same plant is relied on by foresters and land conservationists to clothe denuded slopes, rapidly providing protection from erosion. Because biological control agents are unable to distinguish between plants of a given species in their different roles, the spontaneous appearance in the 1940s in New Zealand of biological control of Manuka was regarded as a mixed blessing.

Although a weed may be of no obvious commercial or aesthetic value, it may nevertheless be an essential link in the interdependence of plants and animals, pests and predators, and its effective control biologically may have unforeseen repercussions. In Puerto Rico, for example, attacks from mole crickets are more severe in the absence of two particular weed species which provide nectar for a predator, and many other such links have been reported (Andres, 1977*a*). Before taking the irrevocable step of introducing a new organism for biological control, therefore, an examination of its full ecological implications must be made.

### 5.7.4 Economic advantages of biological control

Both new pesticides and biological control agents are expensive to develop, evaluate and introduce, but the latter have the advantage that once established they are self-perpetuating and savings continue to be made without extra cost.

Because biological control and its assessment is the result of cooperative effort between many government agencies, accurate costings have seldom been attempted. It is now common to express research costs in terms of scientist years (SY), one SY being the cost of one scientist plus all necessary support costs during 12 months of work. In 1976 this was estimated to be equivalent to £40,000 in North America, and in current programmes costings in terms of SYs should enable a better assessment to be made of the financial advantages of biological control over other methods. Costings that have been made suggest that the cost of introducing a new biological control insect, at 1971 prices, was about £0.5M; this can be compared with an estimate that the successful control of St. John's-wort or Klamath-weed (*Hypericum perforatum*) with the beetle *Chrysolina quadrigemina* was worth over £20M within only 10 years in California alone, due to enhanced land values, reduced need for herbicides and superior weight gains by cattle (Andres, 1977*a* and *b*).

The augmentative method of biological control is one where an organism is multiplied in isolation and released at the most vulnerable growth stage of the weed to augment natural control. This, because it requires annual expenditure, is inherently less profitable than the classical type of control generally used. Control of aquatic weeds with grass carp (*Tillapia* spp.), although it requires annual restocking has been both successful and profitable in California and Arkansas (Andres, 1977*b*).

### 5.7.5 The future for biological control of weeds

Economic and social pressures to reduce reliance on pesticides has encouraged biologists to seek cultural and biological solutions to pest, disease and weed problems. There have been successful local campaigns against individual troublesome species, generally introduced weeds that dominate low quality grazing land for which herbicide treatment would be uneconomic, and also against aquatic weeds. Recently, however, the successful exploitation of a rust fungus in Australia has extended the potential scope for economic control to weeds in other situations, including arable cropping. Despite the many successes, biological control cannot be expected to have more than a small impact on the world's weed problems, and cultivation and herbicides will continue as the basic tools of weed control.

## REFERENCES

AKOBUNDU, I. O. (1980). Live mulch: a new approach to weed control and crop production in the tropics, *Proceedings 1980 British Crop Protection Conference – Weeds*, pp. 377–382

ALTMAN, J. and CAMPBELL, C. L. (1977). Effect of herbicides on plant diseases, *Ann. Rev. Phytopathol.*, **15**, 361–385

ANDRES, L. A. (1977*a*). The biological control of weeds, in *Integrated Control of Weeds* (eds. J. D. Fryer and S. Matsunaka), University of Tokyo Press, Tokyo, p. 262

ANDRES, L. A. (1977*b*). The economics of biological control of weeds, *Aquatic Botany*, **3**, 111–123

ANON, (1978). *Screening Organisms for Biological Control of Weeds*, Commonwealth Agricultural Bureaux, Farnham Royal, Buckinghamshire, p. 6

CRAFTS, A. S. (1975). *Modern Weed Control*, University of California, Berkeley, p. 440

CULLEN, J. M., KABLE, P. F. and CATT, M. (1973). Epidemic spread of a rust imported for biological control, *Nature, Lond.*, **244**, 462–464

CUSSANS, G. W. and WILSON, B. J. (1976). Cultural control, in *Wild Oats in World Agriculture* (ed. D. P. Jones), Agricultural Research Council, London, p. 296

FRYER, J. D. and MAKEPEACE, R. J. (eds.) (1977). *Weed Control Handbook*, vol. 1, *Principles*, Blackwell, Oxford, p. 510

FRYER, J. D. and MATSUNAKA, S. (eds.) (1977). *Integrated Control of Weeds*, University of Tokyo Press, Tokyo, p. 262

HASAN, S. (1974). Recent advances in the use of pathogens as biocontrol agents of weeds, *PANS*, **20**, 437–443

HENSON, I. E. and LITTLE, E. C. S. (1969). Penetration of polyethylene film by the shoots of *Cyperus rotundus, PANS*, **15**, 64–66

KASASIAN, L. (1971). *Weed Control in the Tropics*, Leonard Hill, London, p. 307

NEWMAN, L. H. (1965). *Man and Insects*, Aldus, London, p. 252

NODA, K. (1977). Integrated weed control in rice, in *Integrated Control of Weeds* (eds. J. D. Fryer, and S. Matsunaka), University of Tokyo Press, Tokyo, p. 262

SUTCLIFFE, J. (1977). *Plants and Temperature*, Studies in Biology no. 86, Edward Arnold, London, p. 57

TEMPLETON, G. E., TeBEEST, D. O. and SMITH, R. J. (1979). Biological weed control with mycoherbicides, *Ann. Rev. Phytopathol.*, **17**, 301–10

WAPSHERE, A. J. (1975). A protocol for programmes for biological control of weeds, *PANS*, **21**, 295–303

## FURTHER READING

DIPROSE, M. F. and BENSON, F. A. (1980). The use of high voltage electricity for weed beet control, *Proceedings 1980 British Crop Protection Conference – Weeds*, pp. 545–548

# 6 HERBICIDES

'Chemical weed control is a miracle of our technological age' according to Ashton and Crafts (1973). The very word miracle implies that chemical control of weeds with very small quantities of herbicide is beyond human understanding, and it is certainly true that few general principles of chemical control have been formulated; rather, each use of a particular herbicide for selective weed control in an individual crop is highly specific. The choice of herbicide for a particular situation will depend upon several variables including climate, soil type, prevalent weed species, crop cultivar and method of propagation and management. A herbicide effective and safe for strawberries in North America may not be so in Scotland. In this chapter some aspects of selectivity and mode of action are summarised.

As yet, knowledge and understanding in this field are inadequate to answer most of the questions we ask about how herbicides work. With deeper understanding it may be possible one day to design molecules which will achieve weed control by exploiting the morphological and biochemical differences between plants.

## 6.1 DISCOVERY AND DEVELOPMENT OF HERBICIDES FROM ANTIQUITY TO PRESENT TIMES

### 6.1.1 Weed killing with inorganic compounds

The earliest known account of selective chemical weed control is by Varro in the first century B.C. who refers to the use of 'amurca', the fluid residue left after the extraction of olive oil from olives. Poured around the base of olive trees or 'wherever noxious weeds grow in the fields', the weeds were, he claimed, killed. The herbicidal compound most consistently referred to over the centuries is common salt (sodium chloride) and the phytotoxicity of amurca may well have been due to its salt content. It is claimed that after the sacking of Carthage in 146 B.C. the Romans punished the people by applying salt to their land so that their crops would not thrive, but for some reason there is no record of salt being used for weed control. Later, Arab writers described the use of various sorts of excrement mixed with salt for killing plants, but again apparently not for weed control. In the seventeenth century it was recommended that sea salt be mixed with winter wheat seed, in the proportion of one part salt to two of seed, for selective control of weeds in the seedbed as 'there is nothing which killeth weeds, quicks and other offences of the ground so much as saltness'. However, the margin for error was evidently small, because later it was recommended that, to avoid killing the wheat, the salt should be applied direct to the soil a year in advance, allowing the salt to be leached out of the soil by the winter rains well before sowing the wheat the following autumn. The phytotoxic effects of copper sulphate on wheat seeds were noted in the eighteenth century, when it was frequently used as one ingredient of a cereal seed steep for control of seedborne diseases, but its herbicidal properties remained undiscovered for another century (Smith and Secoy, 1976).

In 1896 Bonnet made the chance observation that Bordeaux Mixture fungicide sprayed on grapes to control downy mildew also blackened and killed the leaves of nearby Charlock (*Sinapis arvensis*). Later he was able to report to the Agricultural Society at Rheims that a 6 per cent solution of copper sulphate (one of the ingredients of Bordeaux Mixture) effectively controlled Charlock growing in cereal crops; the cereal remained unharmed because the spray was shed by the upright, somewhat water-repellent cereal leaves but was retained by the Charlock. So began *selective* chemical weed control which, in this book, is given a broad interpretation and is taken as meaning the use of a herbicide to control weeds without harming the crop in which they are growing, however this is achieved. Ammonium sulphate, copper nitrate and especially sulphuric acid were soon shown to possess herbicidal properties similar to copper sulphate, and further work added Chile saltpetre (sodium nitrate) and calcium cyanamide. Several other inorganic compounds including sodium arsenite, sodium chlorate, borax, ammonium thiocyanate and more recently ammonium sulphamate have been found highly effective as total weedkillers for use on uncropped areas. Sodium arsenite was used selectively beneath some tree crops and it remained for many years the standard method of controlling perennial grasses in rubber plantations. It is also a fast-acting plant desiccant once used for destruction of potato haulm before harvest, but fortunately less poisonous alternatives to sodium arsenite are now available for these purposes. Some inorganic herbicides remain in use; ammonium sulphamate, for example, is recommended for killing the stumps and root systems of felled trees.

### 6.1.2 Organic herbicides

The first indication that organic compounds might become useful herbicides came in 1932 with the discovery of the properties of DNOC, a nitrophenol compound (*Figure 6.1*). Like sulphuric acid, DNOC was

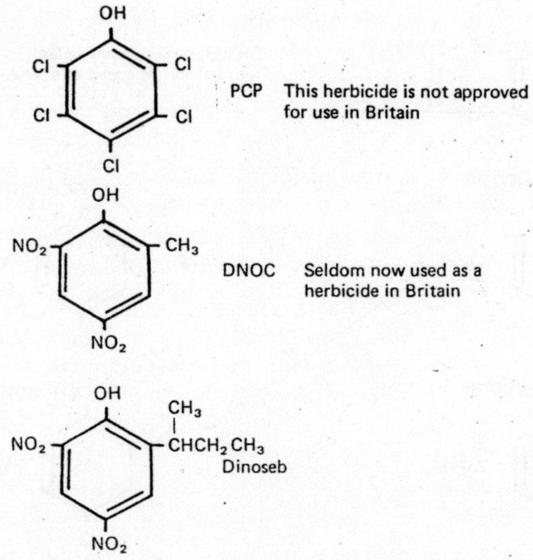


**Figure 6.1** Herbicides derived from phenol.

selective in killing broad-leaved weeds in cereals; in 1946 the related compound dinoseb (*Figure 6.1*) was found to kill most kinds of weeds growing among peas without harming the crop and it remained the standard pea herbicide until superseded by safer and more reliable products twenty years later.

Due in part to ignorance but perhaps mainly because of the non-availability of cheap and reliable spraying machines, neither DNOC nor sulphuric acid were widely adopted by any but the more adventurous farmers. Both substances are toxic and unpleasant and call for skill in mixing and application if accidents are to be avoided; inadequate protective clothing worn by operators when applying DNOC during hot weather has occasionally led to deaths.

The discovery of growth regulator herbicides independently in Britain and the United States during the 1940s was a direct result of research into synthetic derivatives of the natural plant hormone IAA (*Figure 6.2*). Charlock was shown to be controlled selectively in cereals by IAA and further work revealed that the chlorinated phenoxyacetic acid compounds 2,4-D and MCPA (*Figure 6.2*) were even more effective, exhibiting a quite remarkable activity against Charlock and most other broad-leaved weeds at the very low rate of less than 1.0 kg/ha. At about the same time the grass killing properties of propham (*Figure 6.4*) were found in Britain, and 2,4-D, MCPA, propham and the older DNOC were extensively tested during and after the war years. Rapid development was encouraged by the labour shortage and urgent need

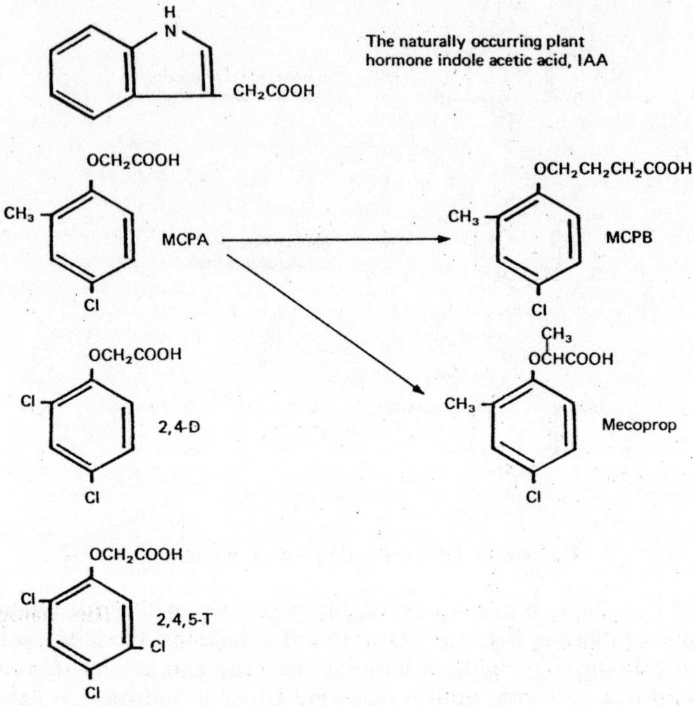


**Figure 6.2** Herbicides structurally related to the natural plant hormone IAA.

for maximum food production; and soon both 2,4-D and MCPA were giving very acceptable weed control in cereal crops when applied with the simple, robust, tractor-drawn sprayers that had been developed. A complete revolution in the management of arable crops had begun.

Following the discovery of these powerful herbicides, and at about the same time of the insecticides DDT and parathion, there began a systematic search for other organic pesticides of all kinds. There are now well over one hundred different herbicides known, of which about eighty are officially approved for sale in Britain. Some, like Bonnet's copper sulphate and 2,4-D are generally absorbed through the leaves. Others are entirely soil absorbed, but many herbicides are able to enter via either leaves or roots without difficulty. Sulphuric acid and other contact herbicides are considered to have a purely local damaging effect on leaves; other herbicides, exemplified by 2,4-D and several newer herbicides (aminotriazole and glyphosate, for example) are translocated via the xylem or phloem to other parts of the plant, although the extent to which this occurs will depend on the stage of growth and on other factors. Some herbicides show marked selectivity for particular crops, while others rely for selective action on a complete separation from susceptible crops in either space or time. By convention, herbicidal applications before the emergence above the soil of either the new shoots of perennial crops or the seedlings of a sown crop are referred to as 'pre-emergence', and applications after the crop shoots appear are termed 'post-emergence'. Herbicides may also be applied before a crop is sown or planted. Application to the entire area is termed 'overall'; less than complete coverage with herbicide can be as a more or less narrow 'band' which usually includes the row of plants, or as a 'spot' treatment of localised patches of weed.

The first organic herbicides were the petroleum or mineral oils, certain types of which possess selective properties; for instance, white spirit, an oil with a boiling range between 140 and 210°C and an aromatic content not exceeding 25 per cent was formerly recommended for weed control in carrots. The cheaper tractor vaporising oil (TVO) was often used instead, but some grades caused crop damage and taint. Urea or triazine herbicides are now preferred to oil for weed

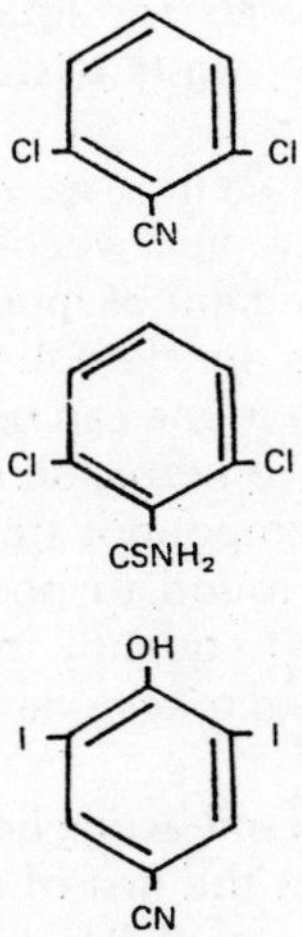


Dichlobenil

Chlorthiamid

Ioxynil

(The similar bromoxynil has bromine atoms instead of iodine)

**Figure 6.3** Three widely used nitrile herbicides. Despite similarity in structure, ioxynil and bromoxynil are foliar contact herbicides but dichlobenil and chlorthiamid are soil acting.

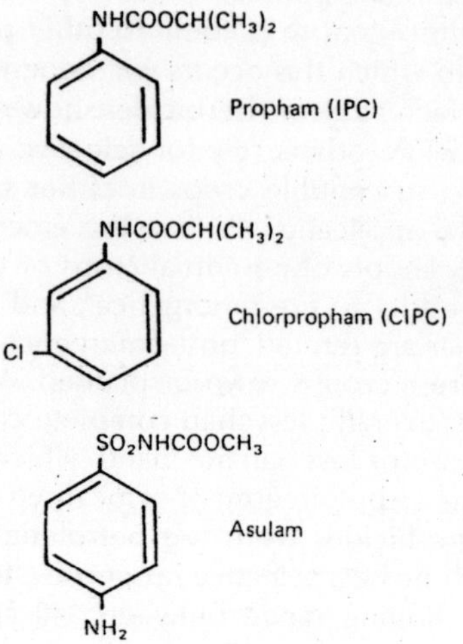


**Figure 6.4** Structures of three carbamate herbicides.

control in carrots. More phytotoxic oils, such as diesel oil, are sometimes mixed ('fortified') with dinoseb for total weed control on uncropped land; sump oil and other waste oils are also effective for this purpose.

The trichloro-substituted phenoxyacetic acid, 2,4,5-T, proved to be more effective than 2,4-D against perennial weeds, particularly woody species, and it forms the main ingredient of most 'brushwood killer' formulations. Two other derivatives are described elsewhere in this chapter; both are non-phytotoxic, but one can be converted to 2,4-D by soil bacteria and another to active herbicide within the tissues of susceptible species. Many other compounds derived from the synthetic plant hormones have been found to possess activity against particular problem species, and are frequently included in herbicide mixtures to increase the range of weed species destroyed.

### 6.1.3 Long-term weed control from soil-acting herbicides

Monuron (or CMU) (*Figure* 6.5) was the first of the substituted urea herbicides, introduced as a total weedkiller but mainly for grass control. The weakness of monuron as a total herbicide is its inability to control deep-rooted perennials, but this has been turned to advantage in its use, together with some other substituted ureas, for control of germinating weed seedlings in deep-rooted perennial crops such as sugar cane, pineapple and asparagus. Diuron (*Figure 6.5*), another urea herbicide, moves into the rooting zone even more slowly than monuron and has been successfully used in some annual crops including cotton. All ureas are root absorbed from the soil, but some, including linuron (*Figure 6.5*) and monolinuron are also leaf absorbed. Uses include weed control in potato, and in a variety of vegetable crops. The members of the urea 'family' show a range of soil persistence and other properties which fit them for weed control in many different

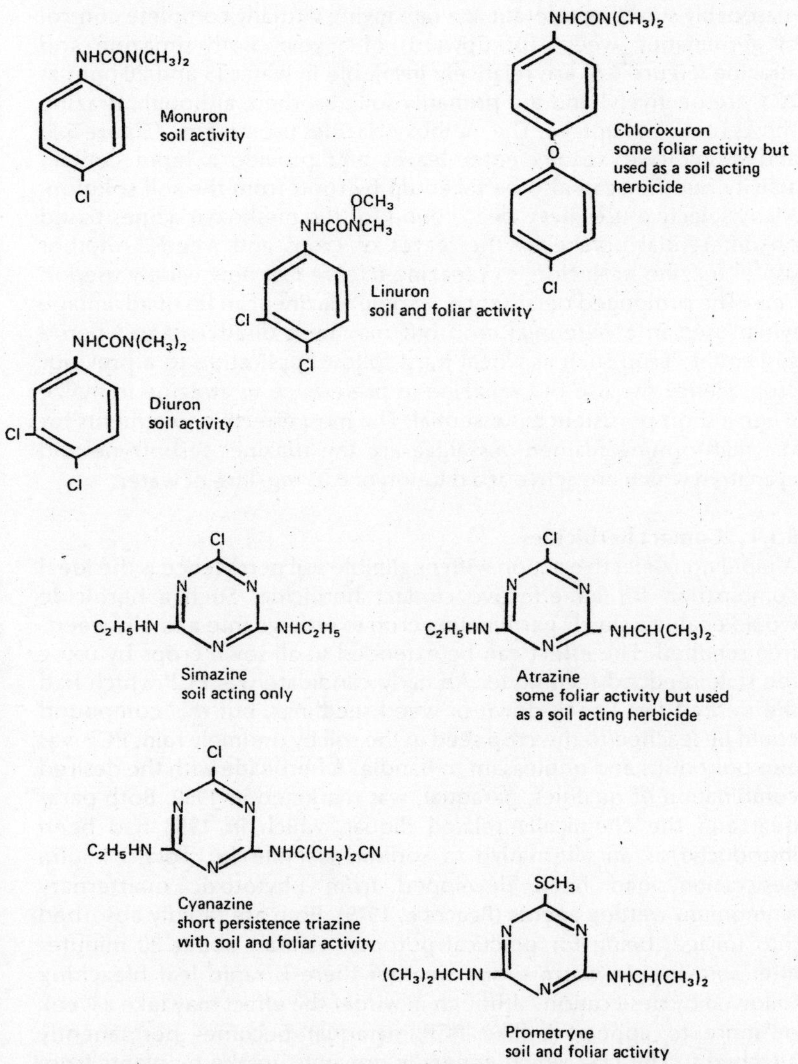


**Figure 6.5** Structures of four urea herbicides and four triazine herbicides.

crops. The more persistent ones such as linuron are sometimes responsible for damage to subsequent crops.

The substituted triazines, introduced from the late 1950s onwards, have probably had a greater effect on horticultural and plantation crop production than any other group of herbicides. First simazine, then atrazine, proved to be highly selective in the maize crop, which is now known to detoxify these herbicides within its tissues. Tree and bush crops and many herbaceous perennials, too, have proved to be reasonably simazine tolerant at a rate giving virtually complete control of germinating weeds for upwards of a year. Both simazine and atrazine (*Figure 6.5*) are relatively insoluble in water (5 and 70 ppm at 20°C, respectively) and are primarily soil absorbed, although atrazine shows some leaf uptake. The methoxytriazines prometryne (*Figure 6.5*) and desmetryne readily enter leaves and provide a rapid contact activity but they can also be taken up by roots from the soil solution. Many selective uses have been found for the methoxytriazines based on differential uptake by the leaves of crops and weeds. Another useful triazine herbicide is cyanazine (*Figure 6.5*) now widely used in peas. The prolonged persistence of some triazines can be of advantage when used in a perennial crop but may be a disadvantage when a susceptible crop such as wheat must follow application to a previous crop. Hence the use of cyanazine in preference to atrazine in maize when a short persistence is essential. The most effective treatments for the mat-forming filamentous algae are the triazines terbutryne and cyanatryn which are active at a dilution of 0.05 mg/litre of water.

### 6.1.4 Contact herbicides

A rapid non-selective action with negligible soil persistence is the ideal combination for an effective contact herbicide. Such a herbicide would enable a slowly germinating crop to emerge into a totally weed-free seedbed. The effect can be extended to all sown crops by using the stale seedbed technique. An early candidate was PCP which had the desired fast knock-down of weed seedlings, but this compound could be leached to the crop seed in the soil by untimely rain. PCP was also poisonous and unpleasant to handle. A herbicide with the desired combination of qualities, paraquat, was marketed in 1960. Both paraquat and the chemically related diquat, which in 1958 had been introduced as an alternative to sodium arsenite for potato haulm desiccation, had been developed from phytotoxic quarternary ammonium wetting agents (Peacock, 1979). Both are rapidly absorbed into foliage, being for practical purposes rainfast about 30 minutes after spraying. In warm sunny weather there is rapid leaf bleaching followed by desiccation, although in winter the effect may take a week or more to appear. Unlike PCP, paraquat becomes permanently attached to soil clay which generally prevents uptake by plants from soil; soil uptake can occur, however, when soils contain no clay. Paraquat is effective against grasses and broad-leaved species, and is formulated alone and in mixtures with soil-acting herbicides. Because paraquat is extremely poisonous to mammals if ingested, great care when handling the concentrate is essential. Diquat, which is much less toxic, is recommended for crop desiccation and for aquatic weed control. Several new techniques, such as direct drilling, have only become practicable since the introduction of paraquat which can destroy existing vegetation rapidly without leaving a toxic residue in the soil (Chapter 8).

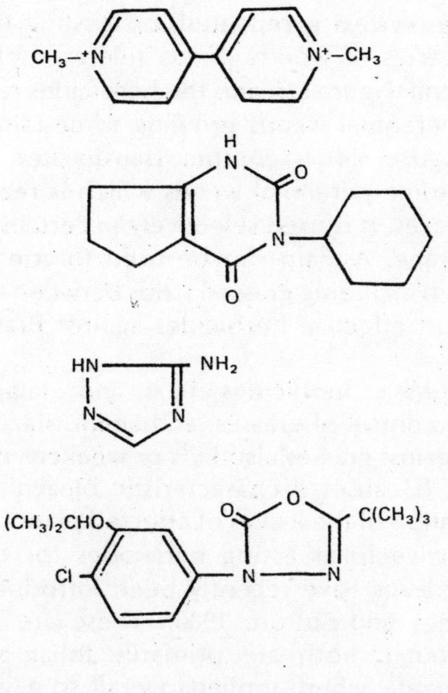


$(OH)_2P(O)CH_2NHCH_2COOH$ Glyphosate

**Figure 6.6** Structures of five important herbicides not chemically related to each other.

### 6.1.5 Herbicides for perennial weeds

To provide permanent control of deep-rooted perennial weeds the herbicide must somehow move from the leaves or roots to the growing points of shoots, roots, rhizomes and other regenerative storage organs. Glyphosate (*Figure* 6.6) is a highly active, non-selective herbicide, absorbed through leaves and translocated to growing points, that is effective in killing annual and perennial species; it is not rapidly broken down in the soil, but because of a combination of low activity via the roots and some adsorption on soil colloids glyphosate is, in practice, foliar acting only. Unlike paraquat, glyphosate is of very low mammalian toxicity.

Typical uses are the control of perennial weeds in plantation crops and nursery stock production and of perennial weeds, especially grasses, in cereal stubble. An entirely different approach to perennial weed control is provided by chlorthiamid and the similar dichlobenil, both of which provide a chemical barrier near the soil surface when applied as granules during the dormant season. Shoots entering this barrier absorb a lethal dose; further shoots grow and as they attempt to reach the surface the process is repeated, exhausting the weed's food reserves. Many species of woody plants tolerate chlorthiamid (*Figure 6.3*) and dichlobenil (*Figure 6.3*) and the herbicides can be used to destroy patches of perennial weeds growing in certain tree and bush crops. Asulam (*Figure 6.4*) is another translocated herbicide effective in controlling some perennial weeds which is readily taken up from either soil or leaves. It is used selectively in certain perennial crops, including sugar cane. Asulam can be used to control Docks (*Rumex* spp.) in grassland, including grassed strips between tree crops, and it is one of the most effective herbicides against Bracken Fern (*Pteridium aquilinum*).

The halogenated aliphatic herbicides TCA and dalapon were introduced for selective control of grasses, and aminotriazole (*Figure 6.6*) which is effective against grasses also kills or weakens most other herbaceous perennials. It causes a characteristic bleaching accompanied by pink discoloration of the leaves of affected plants.

Two remarkably selective foliar acting herbicides for controlling annual and perennial grasses have recently been introduced (Atkin and Wilson, 1980; Finney and Sutton, 1980). These are alloxydim-sodium and fluazifop-butyl; both are primarily foliar acting and appear to be relatively safe when applied overall to a wide range of broad-leaved crops and ornamentals. Although successful against perennial grasses including Common Couch (*Agropyron repens*), some annual species such as Annual Meadow-grass are poorly controlled.

Many other herbicidally active substances, apart from those few already mentioned, are now in general use in crop production (MAFF, 1980). Some, such as the group of herbicides developed to control wild oats in cereal crops, show a high degree of specificity.

### 6.1.6 Herbicides for minor crops

The search for new herbicidal compounds continues, but although there are still several important crops for which no entirely suitable herbicides have been found, the present high cost of meeting the safety requirements of government environmental protection agencies precludes the further development of any pesticide for which a large market is not assured. Thus the relatively small scale production of a wide range of plant species that characterises horticulture must generally wait for a herbicide until a suitable one is found that was developed originally for a large scale crop. The herbicides lenacil (*Figure 6.6*) and phenmedipham, respectively useful as soil-acting residual and post-emergence contact weedkillers in strawberry, were both developed for sugar beet. Lenacil has proved also to be safe for use in a wide range of ornamental perennial species. Future developments including novel methods of application and formulation, and even deliberate selection of herbicide-resistant cultivars, may well increase the range of horticultural situations in which existing well-proved herbicides can be used.

## 6.2 MODE OF ACTION OF HERBICIDES

How do herbicides work? The symptoms of herbicide damage to crops, such as the uncontrolled proliferation of tissues seen with 2,4-D or the grey, withered seedling leaves due to triazines and ureas, are all too familiar. It is harder to elucidate the crucial action or actions of herbicides on plant structure or function that lead inexorably to death. Even now after 30 years of research effort the key biochemical event triggered by 2,4-D remains to be discovered, although it is almost certainly involved with the nucleic acids or cell membranes. However, it may be possible to order the symptoms associated with plant death into primary and secondary effects and so to find the ultimate causes. But there are some effects, such as the stimulation sometimes caused by sublethal doses, that appear to have no direct connection with the lethal action. For example, triazines in low concentration often produce a growth stimulation in crops, and benefits are claimed for deliberate application of very low rates of 2,4-D; the cause of these effects is unknown.

### 6.2.1 Interference with photosynthesis

According to Dodge (1977), almost half of the herbicides approved for use in Britain have a primary effect on photosynthesis, acting either by inhibiting electron transport (including ureas, triazines, uracils and several carbamates) or by diverting the energy produced in photosynthesis (paraquat and diquat). If the light-induced flow of electrons in the chloroplast is prevented by a herbicide, then the supply of ATP (adenosine triphosphate) and $NADPH_2$ (nicotinamide-adenine-dinucleotide-phosphate) will also cease, preventing the uptake of carbon dioxide and the build up of carbohydrates, thereby depriving the plant of metabolites. In the meantime the photosynthetic pigments will continue to absorb light energy; leaf cells have the ability to dissipate some of this energy derived from excitation of these pigments but are probably unable to deal with the amounts generated when electron flow is blocked. It has been suggested that this leads to the destruction of membranes, and thus the loss of cell integrity and death of the leaf tissues.

Paraquat and diquat interfere with photosynthesis by diverting photosynthetic energy from carbon dioxide fixation and their action is more rapid and dramatic than that of the triazines and ureas. In the dark there is little effect, but in the light there is rapid cell destruction as membranes become disrupted, hydrolytic enzymes are released from the vacuole, the pigments are destroyed and leaves begin to desiccate within a few hours of treatment. It is probable that these deleterious processes are initiated by the formation of toxic radicals and peroxides.

The auxin-type herbicides such as 2,4-D, MCPA and 2,4,5-T elicit a range of responses in plants, some of which are fatal, but large plants often outgrow the symptoms and survive. Small traces of these herbicides cause the growing point of many crop species to produce a thin, sometimes twisted stem with deformed leaves; if the dose is sufficiently small, the growing point eventually resumes normal growth, leaving behind a characteristic narrowing of the stem. At the same time that the leading shoot is reduced, these herbicides often cause proliferation of cells in older tissues, manifested as a production of adventitious aerial roots and callus tissue. In sensitive species quite severe symptoms can be caused by small amounts of herbicide that drift from sprayed areas, by herbicide-contaminated sprayers used for other purposes. and even by contact with contaminated clothing.

Other herbicides, including some carbamates and propyzamide, appear to act mainly as mitotic poisons, preventing root and shoot growth by interfering with the normal spindle fibre organisation which is required to enable the chromatids to separate to the daughter cells. The cotyledons of some species of weeds may expand for a time by cell expansion although the seedlings are already dying, because cell division and root growth have ceased.

Respiration is essential in plant and animal tissues to provide both energy in the form of ATP, and the small molecule 'building blocks' required for metabolism and growth. The supply of ATP is regulated by the demand for energy, but when this close link between supply and demand is broken (or uncoupled) then uncontrolled oxidation of substrate will continue until the organism dies through starvation. Many herbicides have some action as uncouplers, but probably only the phenol derivatives, such as dinoseb, DNOC and PCP, kill weeds primarily through this mechanism.

The role of extracuticular waxes will be examined in terms of spray retention and penetration (Sections 6.6 and 6.7); lipids are also required in the formation of the cell membranes. Low rates of TCA, dalapon and several other herbicides are known to reduce extracuticular wax, thereby increasing cuticular transpiration, but this is probably not essential to their toxic action. EPTC and other thiocarbamates not only reduce leaf wax but are also known to inhibit synthesis of the unsaturated fatty acids required in membrane formation, which soon leads to a reduction in the vital functions of cells.

### 6.2.2 Lethal synthesis

There are several cases in which an inherently non-toxic compound is changed to a toxic one within the plant, a process known as lethal synthesis or lethal metabolism. This happens, for example, when the ester and amide forms in which 2,4-D and MCPA are often applied are converted in the plant to the acid. Aminotriazole is probably changed in the plant cells to an even more toxic compound before being detoxified by other plant enzymes; a similar situation is found with the wild oat herbicides flamprop-isopropyl and benzoylprop-ethyl. Perhaps the best understood example of lethal synthesis is the conversion by β-oxidation of the non-herbicidal butyric acid derivatives of 2,4-D and MCPA (known as 2,4-DB and MCPB) back to 2,4-D and MCPA, respectively, in plant tissues. β-Oxidation is an enzymic oxidation of long chain fatty acid molecules which occurs in steps, two carbon atoms being removed from the chain at each step. The removal of two carbon atoms from either 2,4-DB or MCPB, or from butyric acid derivatives of the other 'hormone' herbicides, leaves the active herbicide as a remainder; if, however, an odd number of carbon atoms is added to the acid side chain of the herbicide, then β-oxidation produces a non-herbicidal compound. Wain (1954) reported that when MCPB was applied to Creeping Thistle (*Cirsium arvense*), Charlock (*Sinapis arvensis*) and Annual Nettle (*Urtica urens*) the normal symptoms expected from an application of MCPA were produced, although Clover (*Trifolium repens*) and Celery (*Apium graveolens*) were unaffected by MCPB; all five species, however, were severely affected by the parent acid, MCPA. It is generally accepted that species resistant to butyric acid derivatives are unable to oxidise the side chain fast enough to release lethal quantities of the active acid, although it has been suggested that differences in activity between MCPA and MCPB are at least partly due to relative differences in ease of penetration between the two compounds in resistant and susceptible species (*Figure 6.7*).

Herbicides, especially those that act on photosynthesis, have provided useful tools for plant physiological research that has yielded much of our detailed knowledge of herbicide mode of action. Ultimately it is to be hoped that an increased knowledge of the precise causes of death of susceptible plants and of the basis of selectivity

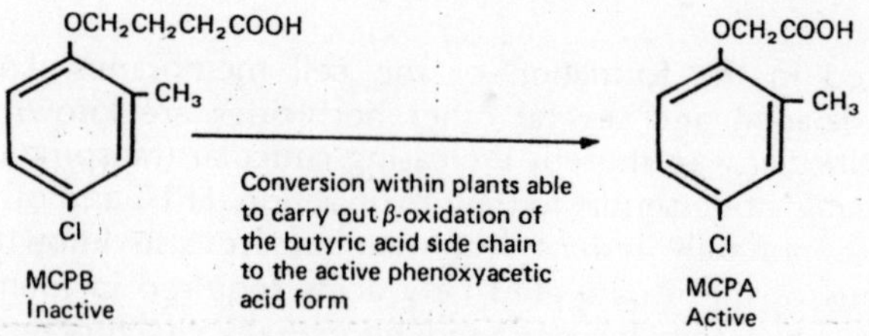


**Figure 6.7** An example of lethal synthesis. Similar changes can be made to 2,4-D and 2,4,5-T, which are then known as 2,4-DB and 2,4,5-TB, respectively.

towards resistant ones may lead to the possibility of actually designing herbicide molecules for specific purposes.

## 6.3 TESTING AND REGISTRATION OF NEW HERBICIDES

Since 1945 screening procedures set up by the chemical industry have produced many compounds showing marked herbicidal activity. Of these some have been rejected because field trials failed to substantiate the early promise shown in the laboratory and greenhouse, some because their use would present unacceptable hazards to users or food consumers, and some for economic reasons, but a few survived to become herbicides we now use. The marketing of a new herbicide is the culmination of several years' testing and development following the initial discovery of activity. In parallel with development for practical use is the accumulation of comprehensive data on all aspects of toxicity, possible accumulation in soil or in food chains, and rates of natural breakdown, to satisfy the stringent requirements of government pesticide registration agencies. Development work of this kind is expensive and if the manufacturer is uncertain of making an eventual profit, further work on the compound may be stopped because he will be unwilling to risk a large outlay on a new compound unless it seems certain to take a substantial share of the herbicide market. Such assurance is doubtful in relation to most horticultural or plantation crops, because of the restricted areas they cover.

### 6.3.1 Finding new herbicides

Major chemical companies produce, in addition to agrochemicals, pharmaceutical and industrial organic compounds, and economies are therefore made by submitting all newly synthesised candidate compounds to as wide a range of tests as possible. The chance discovery of phytotoxic properties in a novel type of organic molecule then leads to the planned testing of related compounds, and thus groups of chemically similar herbicides such as triazines and ureas have been found. Candidate compounds are screened or tested for herbicidal activity using seeds, seedlings or pieces of plant tissues; an effective herbicide screen should ensure the detection of any new and previously unsuspected types of herbicidal activity. The screening procedure adopted by one company has been described by Saggers (1976). In the initial screen, separate batches of seedlings of a range of easily raised monocot and dicot species in families containing major crops and weeds are exposed to foliar-applied and soil-incorporated candidate herbicides. In addition, French Bean (*Phaseolus vulgaris*) seedlings are used to detect growth regulator activity. Qualitative assessments, on a numerical scale, are made of any responses elicited, such as growth inhibition, scorching or growth distortion. Any herbicidal activity may be of interest at this initial stage and therefore rather high rates of compound are applied. Those substances showing activity, probably under 10 per cent of the total, will be considered for further testing using more realistic application rates, a larger range of crop and weed species, and their activity will be compared with that of current herbicides. New substances that continue to show promise after these more demanding tests will be tested further and eventually be subjected to full scale field trials. The chances of finding a marketable pesticide are said to be less than 1 in 12,000 of the compounds tested.

When a novel type of herbicide is found the manufacturer will attempt to ensure his exclusive right of manufacture and sale by taking out patents.

### 6.3.2 Toxicity testing of herbicides

A manufacturer will need to know that a potential herbicide will not present an unreasonable toxic hazard to man before proceeding with expensive development work. Relative toxicities can be compared in terms of the acute oral or dermal $LD_{50}$ for rats (the dose taken by mouth or through the skin that is lethal to half of a group of uniform animals) which is expressed as mg/kg body weight. A low figure indicates high toxicity, a high figure that a relatively large dose is required to kill rats. For example the acute oral $LD_{50}$ in rats for paraquat is between 150 and 200 mg/kg and for simazine is in excess of 5,000 mg/kg, correctly suggesting that paraquat (in the undiluted commercial formulation) presents a greater hazard to the operator than does simazine. But the acute $LD_{50}$ for rats has severe shortcomings as a measure of toxicity in actual use because many other factors are involved (Barnes, 1976). Although rats do appear to respond in much the same way as man to some toxic substances, there are important differences and a relatively low toxicity to rats must never be assumed to apply equally to man. Fortunately most herbicides are of very low acute toxicity, and only paraquat, DNOC and dinoseb (all formulations) are required or recommended to be labelled as poisons in Britain, whereas 29 pesticides of other kinds are considered to be acutely poisonous to man. In many countries those who use highly toxic compounds are required to wear a minimum of protective clothing (see Chapter 7) but reasonable precautions to avoid herbicide contamination of skin and clothing should always be taken, particularly when the concentrated material is being measured and mixed prior to application. More sinister worries for manufacturers and users concern the long-term effects of repeated exposures to sub-lethal doses such as those that might be experienced by farm workers applying the substances or by consumers of food containing traces of herbicides. Manufacturers attempt to use accelerated tests on small animals to assess these risks but it is virtually impossible to prove that any new substance is completely safe. Indeed, using similar tests to those that are applied to candidate pesticides, it has been shown that many natural and generally accepted foods, including cabbage, potato, turnip and milk contain toxins and even carcinogens. Aminotriazole will cause thyroid cancer in a small proportion of rats which have been fed with the compound continuously throughout their lives but some common natural foods are considered to be many times more likely to cause this condition in man than occasional aminotriazole residues in food. Not only the herbicide but its breakdown products in plants, animals and soil must be tested for acute and long-term toxicity, since such products may still be present in crops at harvest or may be leached to drainage water.

### 6.3.3 Registration and approval

Many countries now operate registration schemes for pesticides, with limitations placed on the crops that may be treated and on minimum intervals between treatment and harvest. There are also legal requirements for protective clothing to be worn when applying the more toxic compounds. Such schemes were originally set up to safeguard farm workers, the general public and wild life from needless exposure to pesticides, but consumer groups, alarmed by what they see as environmental contamination now persuade governments to impose increasingly stringent requirements. The discovery that traces of a very highly toxic substance, 2,3,7,8-tetraclorodibenzo-*p*-doxin known as TCDD or dioxin, were present in formulations of 2,4,5-T herbicide has led to widespread public concern and attempts have been made to limit its use. The level of dioxin in the 2,4,5-T used as a component of 'Agent Orange' defoliant during the Vietnam war was up to 500 times greater than the limit of 0.1 mg/kg of 2,4,5-T now recommended by FAO nationally and in Britain by the Ministry of Agriculture, Fisheries and Food, and in addition the rates of application were relatively large; for this reason no meaningful comparison can be made between the Vietnam experience and the peaceful use of the herbicide for conifer release, in rice production and as a brushwood killer. However, it was claimed that in the state of Oregon, women who lived in forested areas in which 2,4,5-T was applied were marginally more likely to suffer miscarriage than other women; on the basis of the evidence, which has since been strongly disputed, a temporary ban on the use of 2,4,5-T for conifer release was made by the Environmental Protection Agency in the United States. Although some countries, including Norway, have restricted the application of this herbicide, West Germany, Australia, New Zealand and Britain have, after thorough investigation, recommended that no restriction should be placed on its use (MAFF, 1979). The lay public have acquired a false conception of great danger inherent in the use of 2,4,5-T, whereas the level of hazard from the trace of dioxin it contains appears to be extremely small compared with many everyday dangers that are universally accepted. A considered judgement has to be made and a calculated risk taken in the light of the available knowledge but there is a danger that unjustified restrictions on the use of agrochemicals will reduce food production.

Another objection to the use of 2,4,5-T, however, is the possibility of taint in drinking water after the herbicide has been applied to water catchment areas. Possible alternative herbicides for woody weed control, such as glyphosate, triclopyr ester and fosamine ammonium, not only cause considerably less taint than 2,4,5-T but are also less toxic (McCavish, 1980).

Several approaches to legislation for the safe use of pesticides have been adopted. In North America, for instance, maximum levels of pesticide contamination, or tolerance levels above which the crop must not be offered for sale have been fixed for particular pesticides in named crops. Similar regulations are now being considered in Western Europe but opponents of the system argue that it fails to discourage contamination up to the legal limit and that in any case such regulations are difficult to enforce.

Britain operates a voluntary two stage system of registration with the PSPS (Pesticide Safety Precautions Scheme) followed by an application for approval to the ACAS (Agricultural Chemicals Approval Scheme). In order that the PSPS may be satisfied that the new product will be safe to use, it requires wide-ranging information on its toxicity, persistence and breakdown products. Once past this hurdle, the new herbicide can be considered for approval by the ACAS, which evaluates its effectiveness for the purposes for which it will be sold. Provided that tests show its performance is comparable or superior to that of existing products then the ACAS and the manufacturer agree on the wording of the instructions given on the label, and the product can be sold as one officially approved. It will then be included in the annually published list of approved products (MAFF, 1980). Such voluntary schemes are said to foster cooperation between government and industry whereas compulsion leads to conflict.

Manufacturers have to satisfy the varied requirements of many different registration authorities in order to sell worldwide and there is need for international collaboration in rationalising the different procedures. Already the EEC countries are moving towards standardised requirements.

### 6.3.4 Nomenclature

Initially a new herbicide will have a code number, but for marketing it will be given one or more trade names. In addition a common name is coined for the active ingredient for all verbal and written non-commercial references to the herbicide. There is general international agreement about approved common names and those approved and published in Britain by the British Standards Institute are, with minor exceptions, similar to those issued by the Weed Science Society of America. However, considerable confusion is caused by the indiscriminate use of both trade names and approved common names by farmers and advisers. A typical difficulty in herbicide nomenclature arises when a mixture of two, three or more active ingredients is marketed; a single easily remembered trade name will in practice be used in preference to a cumbersome list of active ingredients.

## 6.4 FORMULATIONS, SURFACTANTS AND ADDITIVES

### 6.4.1 Formulation and why it is necessary

The first pesticides were either single chemical substances, such as sulphuric acid, or were mixtures freshly made by the user, like Bordeaux Mixture, which is made by mixing hydrated lime with a solution of copper sulphate. Modern organic herbicides in the relatively pure form known as the technical grade are rarely suitable for use straight away, but must first be formulated or processed to change them into a usable form, or formulation. The manufacturer must ensure not only that his product will be effective in the field, but also that it will be reasonably easy and safe to mix and apply and will not deteriorate rapidly in storage. Additives are included in formulations to increase retention and penetration, for inhibiting corrosion and sometimes fire, and even for deterring people from drinking the product! Most formulations are intended, after appropriate dilution, to be sprayed at 200 l/ha or more, but other types are made for much lower volume application, or as granules (Chapter 7). In a broad sense the term formulation is also applied to the particular chemical form of the active principle; for example, the various metal salt, amine and ester forms of 2,4-D and MCPA.

Formulations of readily water soluble herbicides, which present the least problems, are prepared in solid or liquid form with the addition of surfactants and other additives when necessary; they include glyphosate, paraquat, dinoseb, TCA and dalapon amongst others. The commercial formulation of paraquat is liquid and if taken by mouth proves fatal. For amateur use there is a more diluted soluble granular form which prevents accidental ingestion of the highly toxic liquid concentrate.

A solubility in water of about 100 ppm may be sufficient for a finely divided herbicide to be taken up by plants, but it is for practical purposes insoluble in the spray tank. The alternatives are formulation as a wettable powder, which is a finely powdered herbicide with surfactant added to keep the made up spray in suspension, or making an emulsifiable concentrate by dissolving the active ingredient in an oily organic solvent and adding surfactant to the formulation to make, when mixed with water, an oil in water emulsion. Both wettable powder suspensions and emulsifiable concentrate emulsions require agitation in the spray tank to prevent settling out and consequent uneven application. The concentration of active ingredient in wettable powders is usually rounded off to a convenient percentage activity, such as 50, 75 or 80 per cent by addition of clay or talc.

Wettable powders are being superseded by a liquid formulation in which finely divided solid pesticide and surfactant are suspended in water. Known as flowables, or suspension concentrates, these formulations are more easily measured and diluted in the field. (The properties and uses of herbicide granules are described in Chapter 7.)

### 6.4.2 Tank mixes

A grower will sometimes wish to save time and expense by combining two or more pesticides and applying them as a mixture. Pesticide formulations, however, are individually devised for each product, and it cannot be assumed that any two formulations will be compatible. The surfactants used in herbicide and other pesticide formulations are of various types and greasy precipitates occur when some mixtures are made. The chemistry of surfactants and other additives is beyond the scope of this text but observation of a few basic rules should avoid a catastrophe. Any proposed mixture should first be tested by mixing small quantities of the partially diluted herbicide or other pesticide to confirm that there is no reaction or precipitate formed. However, it has been shown that some mixtures reduce the effectiveness of one or other of the components, and therefore the manufacturers should always be consulted before making mixtures on a large scale. An increasing number of properly formulated and well-tested mixtures are now produced which should be used in preference to home-made tank mixes (Chapter 12).

### 6.4.3 Herbicide additives

The user expects modern herbicides to perform well in the field, taking their successes for granted and frequently bewildered by any failures. Since the extent of weed destruction in any particular situation depends upon a complex interaction between plant, chemical and environmental factors, the generally high degree of success with herbicides is remarkable. Most formulations, however, are general purpose ones designed to store well, occupy the minimum space and work equally well on a range of crops. Recent investigations have shown that addition of chemical synergists can substantially increase the effectiveness of several herbicides for specific purposes. For example, it has been reported that the addition of 1.0 per cent mineral oil to a bentazone spray increased the kill of *Chenopodium album* seedlings, whose resistance to this herbicide had been increased by drought, without reducing selectivity to the crop (King and Handley, 1976). Certain mineral oil fractions, such as those used for weed control in carrots, are known to owe their phytotoxicity to solubilisation of the cuticle and adjacent protoplasmic membranes, but the oils generally used as additives possess little inherent toxicity and function indirectly by increasing penetration and uptake. The additional oil will require extra surfactant to create an emulsion at the time of spraying and mixtures of oil and surfactant are available for blending with atrazine, phenmedipham and other herbicides to increase their foliar penetration. Oils can also be added in such a way that the surfactant behaves as a cosolvent for water and oil, creating a clear stable mixture by solubilisation instead of an emulsion of oil droplets in water. Such stable mixtures with glyphosate show enhanced penetration of bark and better kill of woody weeds in contrast to glyphosate alone (Turner and Loader, 1974). Solubilised oils also show advantages with some herbicides when the volume of spray is reduced below 20 l/ha.

Ammonium sulphate at between 1 and 5 per cent w/v increases the herbicidal effectiveness of many herbicides including MCPA, and other related compounds, picloram, aminotriazole, glyphosate and bentazone (Turner and Loader, 1975). Other ammonium compounds and certain organic and inorganic phosphates also show activity as synergistic additives. Curiously little is yet known of the way in which oils and organic and inorganic additives function as synergists, and so far there are few firm recommendations for their use in the field. However, their economic potential is considerable, both by extending the usefulness of existing herbicides and because of the saving made when lower rates of expensive herbicide achieve the same degree of weed control.

### 6.4.4 Antidotes and protectants

The strong herbicide adsorption shown by organic soils prompted attempts to place adsorbents where they would protect plants by preventing herbicide uptake. A pre-planting dip in a slurry of activated charcoal, which adsorbs very strongly, has been used successfully to protect transplants of strawberry, cauliflowers and some conifer species from triazine herbicides. The method has been tried for seeds, too, but the seedling grows away from the protected zone too rapidly for prolonged protection. Residues of herbicide from a previous crop could be adsorbed to a dressing of charcoal, but the amounts required would be too large to make the method feasible.

In 1962 the discovery of specific chemical antidotes to the action of barban on wheat was reported (Hoffman, 1962), and in 1964 naphthyl anhyride (NA) was found to protect maize, sorghum and rice against the effects of thiocarbamate herbicides, such as EPTC. The crop seed is dusted with antidote at 0.5 per cent of the weight of seed before sowing, which then enters the plant and confers an increased tolerance of herbicide. Weed kill is not reduced because weed seedlings do not have access to the NA. The compound N,N-diallyl-dichloroacetamide (R25788) is even more specific and can be applied as a tank mix with herbicide, protecting maize but not reducing weed kill. In maize either NA as a seed dressing or R25788 as a tank mix enabled the safe dose of EPTC to be trebled, at which rate EPTC controlled annual and some perennial grasses. In rice, NA seed treatment allows molinate, which controls red rice (*Oryza rufipogon*), to be used safely. The mode of action of NA and R25788 is not yet known, but they may cause faster herbicide breakdown in the plant (Blair, Parker and Kasasian, 1976). Hoffman foresaw that effective antidotes would allow higher rates of herbicide to be used without crop damage; allow herbicides to be used more safely under adverse weather conditions that might otherwise cause herbicide crop damage; allow herbicides to be used that normally are non-selective in that crop; and lastly give protection to valuable plants that have been accidently sprayed. It is probable that the exciting possibilities revealed by NA and R25788 have encouraged a thorough search for other herbicidal antidotes (Pallos and Casida, 1978).

## 6.5 HERBICIDE RESISTANCE IN WEEDS AND CROPS

Experience with insects and pathogenic fungi over the past thirty years shows that the appearance and build-up of resistance to pesticides in pest populations is almost inevitable. Forms of pest and pathogen capable of bypassing new pesticidal and genetic barriers continually arise, and the chemists and plant breeders find themselves competing in a race to remain at least one move ahead. Abundant evidence of variation in weed response to herbicides has been found, at least some of which is known to be inherited and therefore subject to selection;

yet few instances of genuine resistance have been found in the field. Resistance is often suspected when individual plants or patches of weeds survive a normally effective herbicide treatment, but as there can be many other adequate explanations, such as incomplete coverage with spray, more evidence is required before genuine resistance can be established.

It is thought that the large reserve of non-resistant dormant seeds in the soil, coupled with the fact that the seed shed by any resistant plants would germinate over a period of years rather than all at once, would militate against the emergence of resistance. If resistance is to arise, then it is most likely to do so when there is a combination of limited seed dormancy, minimum soil cultivation and weed species with short life cycles (Harper, 1956). Prediction was made that when resistance did arise it would be in response to the long continued use of one herbicide, and it is therefore not surprising that the repeated annual use of atrazine in maize in North America has produced populations of annual weeds such as Fat-hen (*Chenopodium album*) and Groundsel (*Senecio vulgaris*) resistant to atrazine and some other triazines (Holliday and Putwain, 1977) and ureas (Grignac, 1978). Other cases of weed resistance in the field have been recorded but the use of alternative herbicides has prevented them becoming a serious agronomic problem. It is interesting that most examples of herbicide resistance in weeds appear to be caused by subtle biochemical changes which prevent movement within the plant, usually by conjugation or degradation, or changes within the chloroplast (Souza Machado *et al.*, 1977).

The genetic basis of resistance has been examined and in some cases a simple one or two gene inheritance has been found, rather similar to the situation with insect resistance to insecticides, but in other instances polygenic, quantitative inheritance occurs. Resistance has been found in both inbreeding and outbreeding species; the latter could spread the resistance genes to other areas, whereas inbreeders could perhaps retain resistance in the temporary absence of the selection pressure, the herbicide in question. So far the occurrence and spread of herbicide resistance in weeds has not become a serious problem, but should they become so in the future, then the genetics of resistance will be of vital concern (Gressel, 1978; Holliday *et al.*, 1976).

Crop damage caused by 'selective' herbicides has long been known, for example, grasses are sensitive to 2,4-D and MCPA as small seedlings and when producing flower primordia, although resistant at other times. Unless either visible symptoms or considerable yield reductions are caused, damage from herbicides is likely to go undetected; thus in Australia several recommended and generally used herbicides were shown to reduce cereal yields by about 5 per cent annually over a five-year period (Elliott *et al.*, 1975). It has also been observed that crop cultivars vary in sensitivity to herbicides and recommendations frequently include a clause proscribing use on certain named cultivars. In outbreeding cultivars it may be possible to select for herbicide resistance, and this has been attempted with some success with simazine in oil seed rape (Warwick, 1976). The herbicide metoxuron can be used in most cultivars of winter wheat for control of annual grasses, but the cultivars Hobbit and Maris Huntsman are sensitive. Investigation of the inheritance of sensitivity showed it to depend on one or two recessive genes only, so that eliminating sensitivity in future wheats should be relatively simple. However, in the meantime, the herbicide isoproturon has been found to have a similar weed killing spectrum to metoxuron and is harmless to all cultivars of winter wheat! (Lupton and Oliver, 1976). Metoxuron is also an effective herbicide in the carrot crop, valuable for control of volunteer potato plants, but experience with metoxuron showed a variation from poor to complete control of potato plants in carrot fields. It has now been shown that some potato cultivars tolerate metoxuron, while others are susceptible (Lutman and Davies, 1976). Faulkner (1976) has isolated a paraquat-resistant strain of Perennial Ryegrass (*Lolium perenne*); it is not harmed by paraquat at 0.3 kg/ha, a rate that controls annual grass weeds and is severely damaging to normal ryegrass (Faulkner *et al.*, 1980).

It remains to be seen whether the incorporation of genes for specific herbicide resistance will be practicable, but the prospects are fascinating.

## 6.6 BEHAVIOUR OF HERBICIDES ON PLANT SURFACES

### 6.6.1 Leaf wettability and the retention of sprays

Several obstacles must be overcome before herbicide sprayed on leaves or soil can reach its ultimate destination inside the cells in sufficient quantities to kill the weed. Any one or more of these obstacles may protect the crop from serious damage, enabling a herbicide to be used selectively. The leaf itself provides the first major obstacle for herbicides applied to leaves. Is it a good target for spray droplets, and if it is, then are the drops likely to be retained on its surface long enough for the active ingredient to pass through the next obstacle, the plant cuticle? The upright, narrow leaves of grasses provide a smaller target for spray droplets than the flat leaves of broad-leaved plants, and cereal leaves with a waxy surface retain less spray than the non-waxy leaves of Charlock. The selectivity shown towards cereals and other crops by contact herbicides appears to depend mainly on these simple factors, although with translocated substances there are additional biochemical obstacles which prevent serious herbicide damage within the tissues of resistant species (Holly, 1976).

When the wettability of leaves of Tomato (*Lycopersicon esculentum*) and Pea (*Pisum sativum*) are compared it is obvious that the pea repels water more strongly than the tomato. The ease with which a surface may be wetted by small water droplets is specified by the advancing contact angle which is the angle between the surface and a line tangential to the droplet at its circle of contact when it first wets the leaf (Holloway, 1970). This angle will be large, perhaps 90° or more, on hard-to-wet surfaces such as pea leaf, beeswax or PTFE (non-stick saucepan coating) because the droplet has little contact with the leaf; such droplets will also have a high centre of gravity and be unstable. The advancing contact angle will be correspondingly small for droplets on surfaces that do not repel water because the droplet flattens (see *Figure 6.8*). The water repellency of leaves is modified by a host of

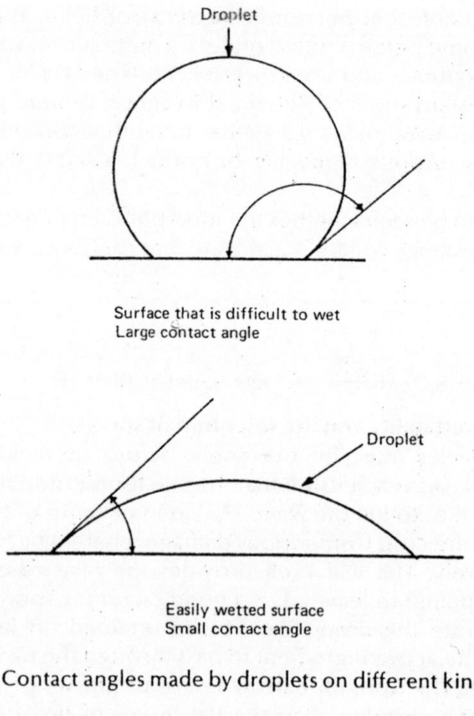


**Figure 6.8** Contact angles made by droplets on different kinds of surface.

factors, in some cases differing significantly between day and night! In general it falls with increasing age of individual leaves and becomes much reduced after weathering in wind and rain, although cotyledons and newly opened leaves are often the most easily wetted. Much of the repellency is due to the form of the extracellular wax that is extruded through the cuticle on to the leaf surface (Martin and Juniper, 1970). There is some doubt about how the wax deposits repel water, since carbon replicas of pea leaves are as effective as leaves, suggesting that the effect is due to the surface form of the deposits as much as to their chemical composition. Certainly the wax is smooth and simple on leaves that do not repel water strongly, whereas on very waxy leaves it is corrugated in a complex manner. Although wax is continuously produced in leaves, the rate depends on the environmental conditions, and greenhouse plants can seldom be used as a satisfactory substitute for plants grown outside in herbicide experiments because the behaviour of the leaves is so different. Another surprising feature of leaf surface wax is that it is markedly reduced by prior uptake of TCA, dalapon and several other herbicides. When droplets flow back over a previously wetted leaf surface the contact angle is often reduced (the receding contact angle), and subsequent wetting and rewetting may reduce the angle still further, although strongly repelling surfaces like pea leaves usually retain this property despite repeated rewettings. All modifications of leaf repellency not only influence the stage of development at which a crop may be sprayed safely, but can also change the degree of weed control obtained from a given rate of herbicide.

### 6.6.2 Other factors influencing retention

The angle of incidence of leaves also affects the fate of droplets arriving on the leaf, and if the leaf is anything but perfectly flat the top-heavy droplets with high advancing contact angle will tend to roll off. As grasses mature some of the leaves flatten and in consequence retain more spray; Hibbett (1969) found that asulam became a better killer of Wild-oat (*Avena fatua*) plants in Linseed (*Lincum usitatissimum*) as the number of flattened leaves increased. Hairs can have a variety of effects on spray retention. At one extreme dense hairiness in some species prevents all contact between water and the leaf, a good example being provided by the specialised hairs on the upper surface of Kariba weed (*Salvinia molesta*) fronds which can open out into a star shape at their tips forming an interlocking water repellent 'double skin' which makes it almost impossible to reach the fronds by spraying (*Figure* 6.9). At the other extreme hairs may serve merely to trap water, making wetting easy, as in the tomato.

Another factor having a marked effect upon leaf retention is droplet size. Work with various sizes of water droplets on pea showed that droplets less than 100$\mu$m diameter were wholly retained, but as the size was increased some of the droplets were shed until at 250$\mu$m all the droplets bounced or rolled off the leaves (Brunskill, 1956) (see

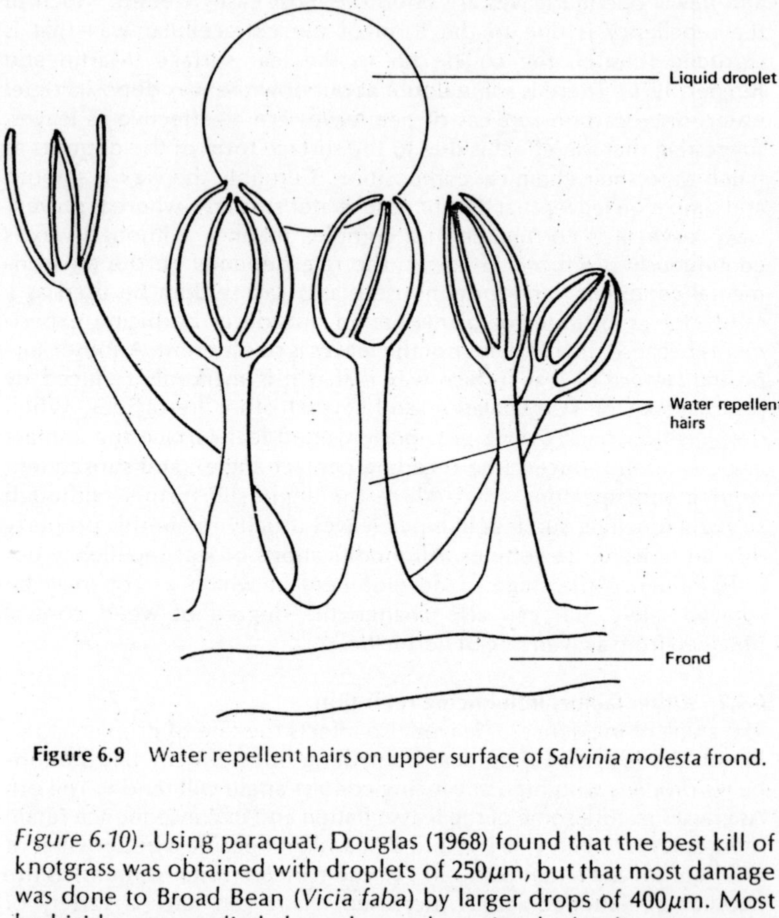


**Figure 6.9** Water repellent hairs on upper surface of *Salvinia molesta* frond.

*Figure 6.10*). Using paraquat, Douglas (1968) found that the best kill of knotgrass was obtained with droplets of 250$\mu$m, but that most damage was done to Broad Bean (*Vicia faba*) by larger drops of 400$\mu$m. Most herbicides are applied through nozzles using hydraulic pressure, a process that inevitably produces a wide spectrum of droplet size. Even if the optimum droplet size for a particular herbicide used were known, this size could be achieved only in a small part of the total spray volume applied when conventional nozzles are used (Chapters 7 and 12).

So far the spray has been assumed to be pure water with a surface tension of about 73$\mu$N/mm but the addition of wetting agents reduces the surface tension, producing profound changes in the behaviour of herbicides on leaf surfaces. The 250$\mu$m droplets of water completely repelled by pea leaves can be made to remain on the leaf by lowering the surface tension to 30$\mu$N/mm (*Figure 6.11*). The addition of sufficient wetting agent to a selective contact herbicide, such as dinoseb,

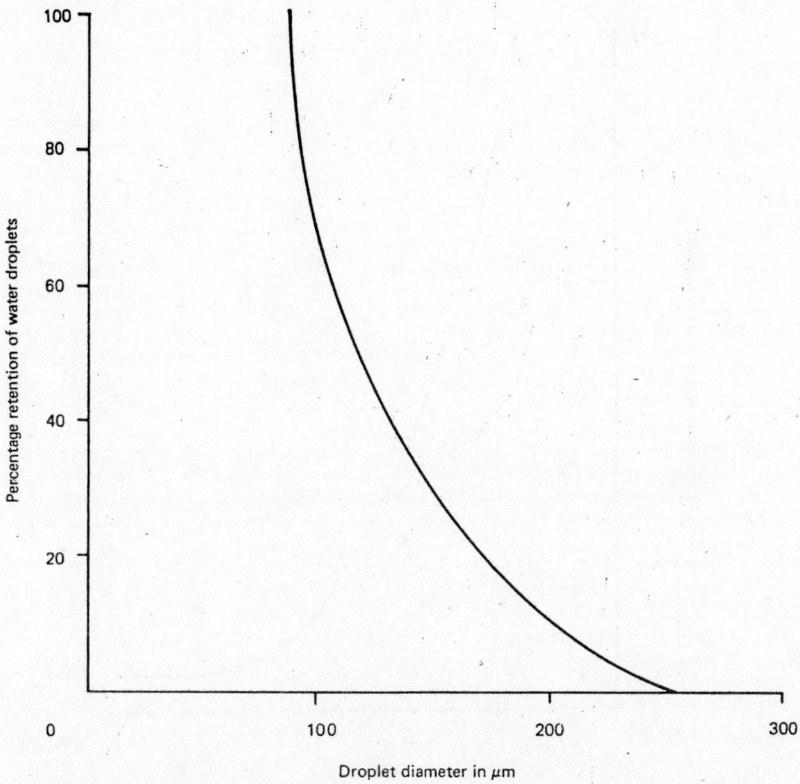


**Figure 6.10** Retention of water droplets on second leaf of young pea plants. (From data of Brunskill, 1956.)

will reduce the surface tension and thereby remove altogether the selectivity shown to peas in its absence. Although spray with a lower surface tension will be retained more on a waxy surface, on an easily wetted surface such a spray will tend to coalesce into a continuous film and part of the spray will then run off the edge of the leaf. The concentration of wetting agent (and hence the surface tension) in foliar applied sprays must be controlled to obtain the desired relative wetting of crops and weeds if crop damage and poor weed control are to be avoided. An appropriate amount of wetting agent will have been included in commercial formulations and extra wetter should not be added unless specified by the manufacturer.

Retention on plants is also affected by the weather, and rain falling soon after application will wash some of the deposit off and redistribute some to other parts of the plant. If the latter are more

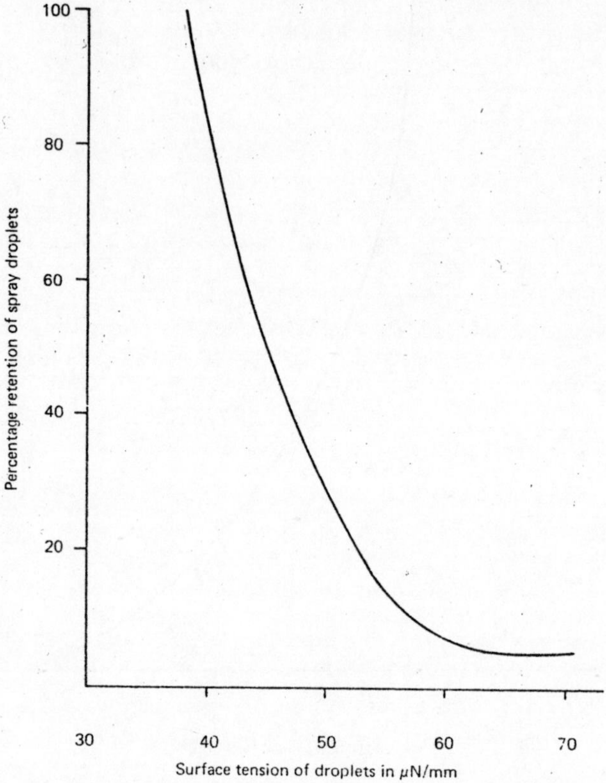


**Figure 6.11** The effect of surface tension on the retention of 250μm droplets on pea leaves. The surface tension was reduced by adding methanol. (From data of Brunskill, 1956.)

responsive sites for uptake, then the loss of herbicide will be offset and even a substantial loss may fail to reduce effectiveness.

## 6.7 FATE OF HERBICIDES IN PLANTS: UPTAKE, MOVEMENT AND METABOLISM

### 6.7.1 The cuticle and surface wax

Before they can work, herbicides must somehow enter the plant. The cuticle, which envelops the aerial part of plants, might be expected to act as an obstacle to entry, but it turns out to be surprisingly permeable. For example, it is common practice to correct micronutrient deficiencies by spraying the appropriate nutrient on the foliage.

Experience has also shown that a great variety of herbicides and systemic fungicides and insecticides can also enter the plant this way. Young roots, too, are covered by a cuticle, but this does not appear to hinder two-way movement of water and mineral ions, and as with leaves many exotic organic substances can enter freely.

The cuticle enables plants to survive in a hostile dry environment by preventing loss of water and discouraging the entrance of parasitic fungi and bacteria. It is a continuous non-cellular membrane made up of wax platelets enmeshed in a matrix containing long chain alcohols, esters, fatty acids, alkanes, ketones and aldehydes (Bukovac, 1976), and since it contains both polar and non-polar groups it combines hydrophilic and hydrophobic properties. When wet, the normal condition at sunrise or after rain, it becomes swollen, and in this condition water and other substances move through with relative ease. For this reason rewetting of the spray deposit will aid entry, and the use of glycerine to delay drying has been claimed to increase penetration. It is now common to apply glyphosate in a gel, with similar advantages (Chapter 7). When the extracellular wax is lightly brushed away not only is wettability improved and water loss (cuticular transpiration) increased but applied chemicals penetrate more easily, suggesting that the extracellular wax is an even greater obstacle to entry than the cuticle itself. As might be expected, cuticles are easily and quickly penetrated by oils and by lipid soluble substances, a finding that led some workers to propose a separate lipoidal pathway through the cuticle; no such pathway, however, has ever been detected. Franke (1967) claimed to have found pathways through the underlying cell walls almost reaching to the cuticle. He demonstrated these after treatment with mercuric chloride, and his electron microscope photographs show them to be clustered near trichome (hair) bases, near guard cells and adjacent to underlying cell walls. These now appear to be artifacts, caused by the mercuric chloride following invisible channels in the cuticle! The significance of this finding for pesticide entry is unknown.

Other attributes of both the plant surface and the spray influence the ease of cuticular penetration, and when herbicidal selectivity is involved these are often highly specific to a crop species or cultivar. The extracellular wax on different parts of the leaf surface varies in kind and properties, usually being more easily penetrated on the lower side and adjacent to guard cells and trichomes (hairs). The conditions of growth, including temperature, relative humidity and amount of sunlight all interact to modify the amount and kind of wax extruded, producing corresponding changes in herbicide retention and penetration. The acidity, concentration of active ingredient and amount of surfactant (surface active agent or wetter) of the spray droplets will all influence penetration independently of any effect on retention. When the concentration of wetter is raised above the critical micelle concentration (generally between 0.01 and 0.5 per cent) there is no further reduction in surface tension and no increase in wetting of water-repellent leaves. However, some herbicides, such as difenzoquat, a selective wild-oat killer in cereals, benefit from a substantial further increase in surfactant concentration by achieving enhanced penetration (Merritt, 1980).

### 6.7.2 Contact and translocated herbicides

The plant may be considered in terms of two closely intermeshed transport systems, the apoplast, or dead part, comprising all cell walls, intercellular spaces and xylem, through which the bulk upward movement of transpired water passes, and the symplast made up of all the cytoplasm of living cells and sieve tubes interconnected by the plasmodesmata. These concepts have enabled an overall view to be taken of the movement of water and solutes within the plant without becoming too involved with the rather poorly understood details of the mechanism of movement within certain parts (especially of the symplast).

Contact herbicides have an essentially local destructive action upon cuticle, cell walls and underlying membranes and cytoplasm; if damage is sufficiently widespread then small annual plants will be killed. However, contact herbicides can become dispersed, and traces of diquat used for haulm desiccation in potatoes during drought conditions have occasionally been detected in the tubers. Paraquat and diquat are for practical purposes contact herbicides, but when the rate of application is low, and especially when several hours of darkness follow application, paraquat is exported, it is thought, via a reverse flow of the xylem (apoplast) to other parts of the plant; application to one leaf may be sufficient to kill all the aerial growth.

But what of the many translocated herbicides applied to roots and foliage? Transport from the roots of soil-applied pesticides is predominantly in the apoplast system, although the casparian strips in the endodermis ensure that all substances moving from the relatively permeable root cortex cells to the stele must pass through the protoplasm of the endodermis. Triazines and ureas and most other soil acting materials travel up the apoplast system (xylem) and accumulate in leaves where they penetrate the cytoplasm and proceed to disrupt photosynthesis (see below). However, only a fraction of the herbicide that enters the root arrives at the shoot.

Many of the early data on herbicide movement in plants were obtained using radio-labelled herbicides, on the assumption that at least for some days the radio-label would indicate the whereabouts of the herbicide, but it is now realised that some forms of neutralisation begin as soon as the herbicide enters. Plants possess methods of neutralising herbicides by oxidation, reduction, deamination, hydrolysis or conjugation, thereby preventing them reaching a sensitive area. Juice extracted from maize tissues for example, will detoxify simazine and atrazine by hydroxylation and in consequence maize, sorghum and sugar cane (and some related grass weeds) are immune to them. The substance responsible is benzooxazinone which is present throughout the maize plant although the most active neutralisation occurs in the roots. This is only part of the story, however, as there are susceptible species capable of hydroxylation (wheat and rye) and strongly resistant ones that are not; maize is also known to be capable of 'fixing' simazine by conjugation into a peptide compound. Ureas, too, are metabolised by many species, but apparently there is no obvious link between species susceptibility and breakdown. Neither ureas nor triazines are moved in the symplast, but move to leaf margins in the apoplast where they cause the characteristic progressive marginal wilting and grey discoloration.

The evidence on movement of phenoxyacetic acid derivatives (2,4-D, MCPA, 2,4,5-T) is contradictory, but movement to the roots is certainly small, though sometimes sufficient to damage roots (Tottman and Davies, 1978) and rhizomes. Rather, what little export there is from leaves appears to be mainly upwards with the transpiration stream in the xylem to the younger leaves. These herbicides are rapidly neutralised in plants, by degradation of the acetic acid side chain, hydroxylation of the aromatic ring and formation of conjugation products. However, none of these processes has yet been convincingly demonstrated as a plausible basis of selectivity for hormone type weedkillers.

The compound 2,3,6-TBA, recommended for use in mixtures based on MCPA for control of difficult weeds in cereals, is exceptionally stable in plants, and residues of this herbicide can often be recovered from the straw. Glasshouse tomatoes and cucumbers grown on contaminated straw can be severely damaged. On the other hand some volatile herbicides are lost in the same way as water during transpiration and it has been estimated that 90 per cent of the dichlobenil that enters the plant is lost in this way; the remainder forms conjugation products and is metabolised in species that tolerate this herbicide.

A recent review of the metabolism of herbicides in plants (Naylor, 1976) concluded that our information is still in a rudimentary state. The current official requirement for detailed information on herbicide breakdown products in plants (and in other parts of the environment) should eventually correct this deficiency. However, despite over thirty years work on all aspects of the behaviour in plants of the phenoxyacetic acid derivatives, no clear picture of the relationship of this behaviour to crop selectivity has yet emerged.

## 6.8 HERBICIDES IN THE SOIL: ADSORPTION, METABOLISM AND FATE

### 6.8.1 Soil factors

Pesticides of all kinds find their way to the soil. Some are deliberately added to control weeds and soil pests, but in addition smaller amounts of materials originally applied to leaves for weed, insect and disease control arrive there. Some foliar spray runs off, or is washed off by rain, and a proportion misses the intended target altogether. The soil provides a dynamic complex of interacting chemical, physical and biological systems, and once in the soil a pesticide becomes an integral part of this complex. Herbicides will be permanently lost from the soil by plant uptake, biological decomposition, physical degradation and photodecomposition, leaching to drainage water and by very long-term adsorption to soil colloids (*Figure 6.12*). Substances with a high vapour pressure will be lost by volatilisation, especially if on or near the surface (*Figure 6.13*).

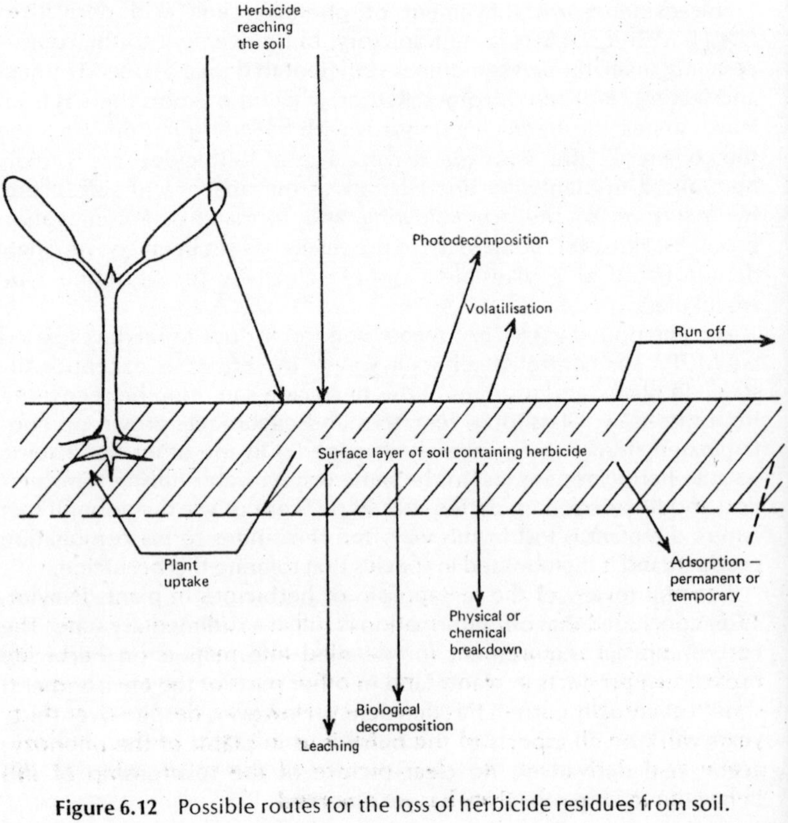


**Figure 6.12** Possible routes for the loss of herbicide residues from soil.

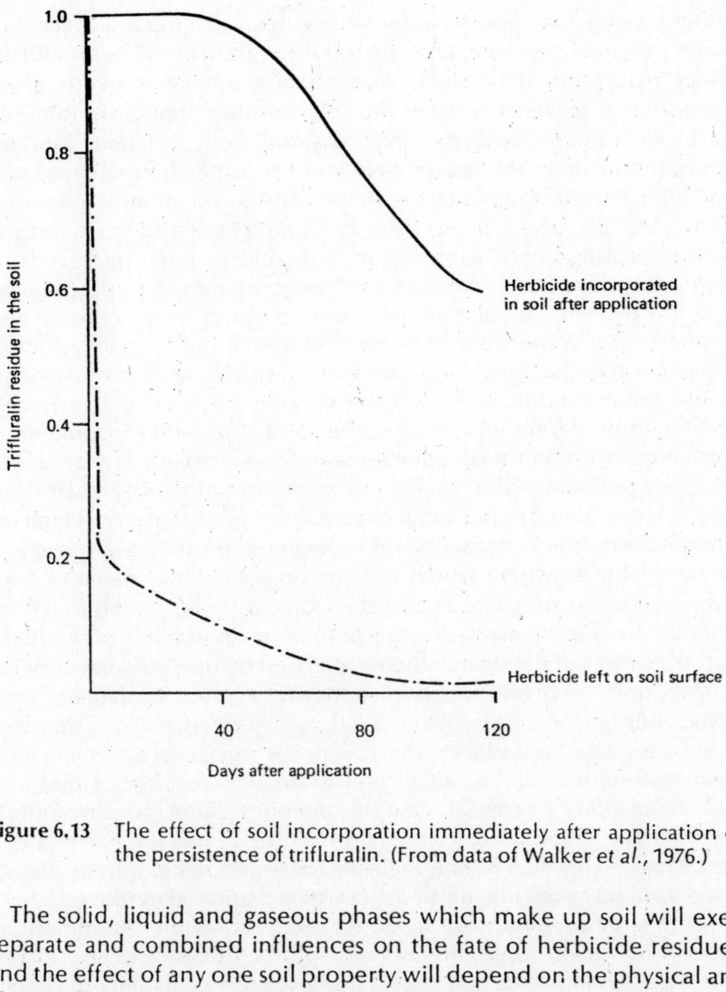


**Figure 6.13** The effect of soil incorporation immediately after application on the persistence of trifluralin. (From data of Walker *et al.*, 1976.)

The solid, liquid and gaseous phases which make up soil will exert separate and combined influences on the fate of herbicide residues, and the effect of any one soil property will depend on the physical and chemical properties of the herbicide. The solid phase is predominantly mineral in origin, except in organic soils, and part of it is usually very finely divided. Mineral particles less than 0.002 mm diameter, the clay fraction, have extensive surfaces (over 2 $m^2$ per g) and clays have, in addition, a layered structure with large internal surfaces. The decomposed organic fraction, too, has similar colloidal properties, which also offer an immense area for adsorption. From the plant's point of view an essential component of soil is a dilute solution of the mineral ions necessary for growth, known as the soil solution. This solution has a water potential of about –0.3 bar at field capacity, offering little resistance to uptake but, due largely to the very strong adherence of the dipolar water molecules to mineral surfaces, the water potential rises steeply as the soil dries, until at –15 bar no further water absorption is possible. As herbicide uptake is nearly always proportional to water uptake, the soil moisture status will influence herbicide uptake strongly. Well-drained soils contain modified atmospheric air in the spaces between the water-film-covered solid particles. The air is in contact with the outside via an interconnecting network of air spaces, but as these become longer and more tortuous with increasing depth so diffusion is unable to keep pace with the respiratory demands of the roots and microorganisms. Carbon dioxide will rise from its 'usual' 0.03 per cent to about 1 per cent in well-drained soils. Water held in heavy soils above field capacity reduces the space available for air and hence the ease of gaseous exchange and under these conditions the carbon dioxide can rise to 10 per cent. Linking these different phases are the plant roots and a population of microorganisms and small animals, which has been said to be 'a large diversive population often living in a ravenous state of near starvation'. The microorganisms produce enzymes in great variety which are liberated into the soil, capable of inducing and catalysing changes in any available substrate which can provide fractions usable as food, even when the substrate is adsorbed on soil colloids. Although the microbial component of the soil population is starved of a suitable substrate most of the time, it is characterised by bursts of activity when a fresh supply of organic material suddenly becomes available or when some restraint on activity is removed, as when frozen soils thaw or a dry soil is wetted. Conditions that favour soil microbial activity, namely warmth, moisture and a suitable substrate, will also favour herbicide decomposition. A special case of decomposition is provided by 2,4-DES. This compound is itself non-herbicidal, but soil bacteria convert it to 2,4-D, which is then available for root uptake; thus its use is as a soil-applied herbicide unlike 2,4-D which is applied to foliage.

Some allowance must be made for soil type when recommending rates of herbicide application because of the influence of soil adsorptive capacity and other soil factors on herbicide availability to roots; a rate required to control weeds on heavier soils may cause crop damage on very light ones and conversely a rate that is effective yet safe for crops on a light soil may be insufficient in heavy ones. With simazine, for example, the proportion of herbicide adsorbed during a standard adsorption test can be more than three times as great on medium and heavy soils as on sands. Heavy soils not only have more clay than lighter ones but tend also to contain more organic matter. In Britain a simple classification of soil types, based on the 'feel' of moist soil between the thumb and fingers, is now used in both official and manufacturers' recommendations for soil applied herbicides (Eagle, 1976; Fryer and Makepeace, 1978). Other schemes of soil texture classification, such as that used by the United States Department of Agriculture, may fail to provide a valid basis for predicting an appropriate rate of application (Eagle, 1976).

### 6.8.2 Herbicide factors

The properties of the herbicide such as water solubility, strength of adsorption to colloids, tendency to form charged ions, volatility and chemical stability, will interact with those of the soil to determine the fate of the herbicide. In the past, water solubility has often been quoted as though it alone were a yardstick indicating relative speed of movement through the soil, but this is misleading and no single factor should be considered in isolation. It might be thought that the low solubility of simazine and lenacil, 5 and 6 ppm at 20°C, respectively, would inhibit downward movement in the soil. But it has been estimated that simazine applied at 1 kg/ha could dissolve completely in less than 3.0 cm rainfall. Dry conditions after spraying do reduce the herbicidal effect of both simazine and lenacil, but once the herbicide has entered the soil following rain, an equilibrium between adsorption and desorption determine any subsequent movement. Some herbicides, such as picloram, fenuron, TCA and dalapon, are both readily soluble and not strongly adsorbed and so easily become leached deeper into the soil profile.

The neutral molecules of many herbicides readily adsorb hydrogen ions to one of the nitrogen atoms thereby becoming positively charged. Such charged particles are strongly attracted to and adsorbed onto the soil colloids, which are mainly negatively charged at normal pH values; this behaviour is typical of the triazines, ureas, uracils and paraquat and diquat. Those without the tendency to become positively charged, such as dalapon, MCPA and dicamba are not attracted and may even be repelled by soil colloids. It is found that each kind of herbicide partitions in any particular soil so that the ratio between concentration of herbicide in the soil solution and the concentration adsorbed is approximately constant. In the extreme case the herbicide is so strongly held by the colloids that none remains in solution, and this is the state in which paraquat and diquat exist in soil, securely bound to clays (especially montmorillonite). More usually an equilibrium is established between the liquid and the solid phase, such that an increase in soil water following rain will dilute the herbicide solution and lead to desorption from the solid phase. Plant uptake will have the same effect, but diffusion of herbicide is so slow in solution that growth of roots into fresh soil zones is a more effective way of maintaining a herbicide supply to the plant roots. Both clay and organic matter are involved in herbicide adsorption, but for most soil-applied herbicides organic matter adsorption predominates. In highly organic soils, not only will the herbicide have difficulty moving into the soil from the surface but at equilibrium little of it will be desorbed and present in the soil solution. These problems can be partially overcome by applying higher rates followed by incorporation of the herbicide into the soil. The vapour pressure of most soil-acting herbicides is so low that loss due to volatility is normally neglible. However, under exceptional circumstances substances with low vapour pressure can become vaporised; for instance, it has been claimed that simazine inadvertently sprayed on hot water pipes in a glasshouse produced severe leaf damage in seedlings within 24 hours. The more volatile herbicides such as EPTC are rapidly lost to the atmosphere unless they are either well incorporated in dry soil or washed and sealed in by rain or irrigation (*Figure 6.13*). Chlorthiamid and dichlobenil are normally applied as granules which reduces vapour loss until there is sufficient rain to carry the herbicide into the soil. Once there, adsorption is rapid and loss by evaporation is much reduced. Heavy rainfall tends to prevent loss of such volatile herbicides by providing a water filled barrier, but repeated light rain actually increases volatilisation because the herbicide is not carried into the soil and the water competes with herbicide for adsorption sites. The importance of chemical degradation of soil-applied herbicides in non-biological soil systems is hard to assess, especially as the same physical conditions, warmth and available water, generally favour both chemical and biological breakdown (*Table 6.1*). It was assumed that most degradation was due to microorganisms but there is increasing evidence that some herbicides, triazines for example, are mainly degraded by non-biological systems in the soil (Crosby, 1976).

The effects of weather and soil type on herbicide persistence and the importance of residues on subsequent cropping are discussed by Walker (1977).

*Table 6.1*

*The rate of breakdown of propyzyamide, simazine and linuron in sandy loam soil held at three temperatures, kept either wet or dry. The figures show the number of days before the level of herbicide residue fell to half of that originally applied, a value known as the 'half-life'. (From Walker, 1977.)*

| | *Half-life of herbicide in days* | | | | | |
|---|---|---|---|---|---|---|
| *Herbicide* | 10°C wet | 10°C dry | 20°C wet | 20°C dry | 25°C dry | 30°C dry |
| Propyzyamide | 110 | 350 | 38 | 121 | 73 | 44 |
| Simazine | 120 | 454 | 49 | 186 | 122 | 80 |
| Linuron | 116 | 232 | 75 | 150 | 122 | 100 |

### 6.8.3 Accumulation of herbicide residues

Simazine was shown in the early 1960s to provide control of germinating weeds throughout the growing season. If repeated annual applications were made would it accumulate? Fears of such a build up prompted thorough investigations in many countries, and although answers to some of the questions asked have been found, the complex and dynamic nature of the soil environment poses its own peculiar difficulties for investigators. In general, most such fears have proved groundless. Both scientific investigation and field experience over many years have shown a rapid rate of decline in soil residues. Where the animal loss from all causes exceeds 80 per cent, as it almost invariably does with the herbicides in common use, then build up will be negligible (*Figure 6.14*); even a very low rate of fall of only 50 per cent per annum will theoretically lead to a maximum accumulation of about twice the annual rate of application. In practice, too, it is likely that a rotation of different herbicides would be used during the life span of a long-term crop. Existing pesticide residues in soil may reduce the rate of degradation of a herbicide, for example, residues of carbaryl, a carbamate insecticide, will prolong the life in the soil of carbamate herbicides such as chlorpropham. Such effects may be important in the short term, but have negligible effect on year to year accumulation.

Using the knowledge gained from experiments under controlled conditions combined with weather data, computer predictions of probable herbicide breakdown rates can now be made. These can indicate when too rapid a decline in soil herbicide residue since application may predispose to particular weed problems, or when, after too slow a breakdown, some crop damage can be expected.

### 6.8.4 Selectivity and soil-applied herbicides

How do soil-applied herbicides achieve useful selectivity in so many kinds of crops? Some cultivars possess the ability to detoxify some herbicides once they have entered the plant, and if the process is fast enough the plant may escape damage. But in cases where there is limited intrinsic resistance, selectivity is reinforced by separating the crop roots from the herbicide as completely as possible. In the Broad Bean (*Vicia faba*), for example, crop damage from an application of simazine can be avoided provided that bean seeds are sown at least 7.5 cm deep. This so-called depth-protection probably accounts for much of the observed selectivity with many crops, although even the broad bean possesses some tolerance of simazine. At one time it was assumed that herbicide absorption occurred entirely through the roots, but although this is the principal point of entry for most soil-applied herbicides it is now established that small but significant amounts also enter the young seedling stems as they break through

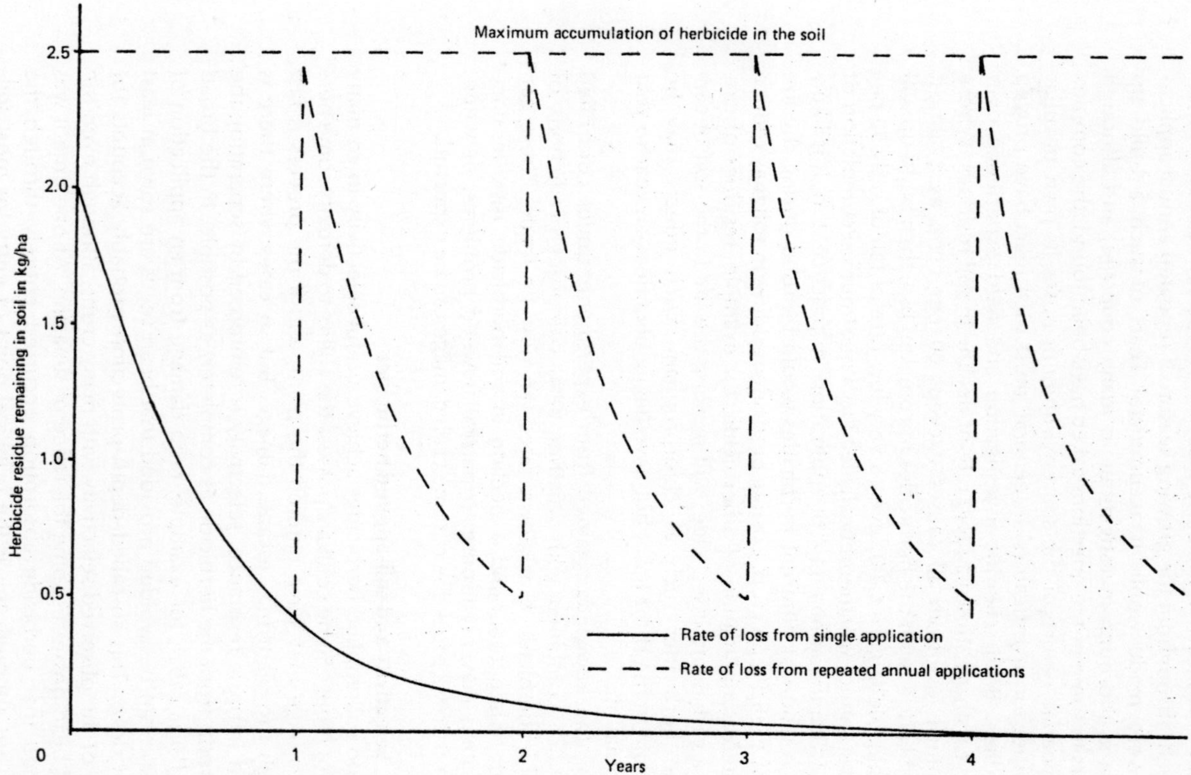


**Figure 6.14** Theoretical curves showing the loss from a single application of 2 kg/ha (solid line) and from repeated annual applications of 2 kg/ha (broken line) when the annual rate of loss is 80 per cent of the amount applied. (From Sheets and Kaufman, 1970.)

the herbicide layer. This is especially true of the volatile thiocarbamate herbicide EPTC which forms a contaminated layer near the soil surface in which the herbicide is either adsorbed to colloids, in solution, or present as vapour. As grass seedlings push through this zone the coleoptile can absorb a lethal dose; the same soil concentration of EPTC can be shown to have little effect when available for root uptake only. It is suggested that selectivity may be achieved by the growing point of the crop either remaining below the contaminated zone until the herbicide has dispersed or else by moving up through it too rapidly for much absorption to occur. Uptake of oxadiazon, too, seems to be confined to the emerging weed shoots as they push through the thin surface layer of soil containing the herbicide. It has to be admitted, however, that often the mechanism of selectivity of soil-applied herbicides is not fully understood.

## REFERENCES

ASHTON, F. M. and CRAFTS, A. S. (1973). *Mode of Action of Herbicides*, Wiley, New York, p. 504

ATKIN, J. C. and WILSON, C. W. (1980). Control of couch grass (*Agropyron repens*) in an arable rotation with alloxydim sodium, *Proceedings 1980 British Crop Protection Conference – Weeds*, pp. 437–443

BARNES, J. M. (1976). Toxic hazards in the use of herbicides, in *Herbicides* (ed. L. J. Audus), vol. 2, Academic Press, London, p. 564

BLAIR, A. M., PARKER, C. and KASASIAN, L. (1976). Herbicide protectants and antidotes – a review, *PANS*, **22**, 65–74

BRUNSKILL, R. T. (1956). Physical factors affecting the retention of spray droplets on leaf surfaces, *Proceedings 3rd British Weed Control Conference*, 593–603

BUKOVAC, M. J. (1976). Herbicide entry into plants, in *Herbicides* (ed. L. J. Audus), vol. 1, Academic Press, London, p. 608

CROSBY, D. G. (1976). Non-biological degradation of herbicides in the soil, in *Herbicides* (ed. L. J. Audus), vol. 2, Academic Press, London, p. 564

DODGE, A. D. (1977). Some mechanisms of herbicide action, *Sci. Prog., Oxf.*, **62**, 447–466

DOUGLAS, G. (1968). The influence of size of spray droplets on the herbicidal activity of diquat and paraquat, *Weed Res.*, **8**, 205–212

EAGLE, D. J. (1976). Soil texture classification for the adjustment of herbicide dose, *Proceedings 1976 British Crop Protection Conference – Weeds*, pp. 981–988

ELLIOTT, B. R., LUMB, J. M., REEVES, T. G. and TELFORD, T. E. (1975). Yield losses in weed-free wheat and barley due to post-emergence herbicides, *Weed Res.*, **15**, 107–111

FAULKNER, J. S. (1976). A paraquat resistant variety of *Lolium perenne* under field conditions, *Proceedings 13th British Weed Control Conference*, pp. 485–490

FAULKNER, J. S., LAMBE, C. B. and HARVEY, B. M. R. (1980). Towards an understanding of paraquat tolerance in *Lolium perenne*, *Proceedings 1980 British Crop Protection Conference – Weeds*, pp. 445–452

FINNEY, J. R. and SUTTON, P. B. (1980). Planned grass weed control with fluazifop-butyl in broad-leaved crops, *Proceedings 1980 British Crop Protection Conference – Weeds*, pp. 429–436

FRANKE, W. (1967). Mechanisms of foliar penetration of solutions, *Ann. Rev. Pl. Physiol.*, **18**, 281–300

FRYER, J. D. and MAKEPEACE, R. J. (1978). *Weed Control Handbook*, vol. II, *Recommendations*, Blackwell, Oxford, p. 532

GRESSEL, J. (1978). Factors influencing the selection of herbicide resistant biotypes of weeds, *Outlook on Agriculture*, **9**, 283–287

GRIGNAC, P. (1978). The evolution of resistance to herbicides in weedy species, *Agro-Ecosystems*, **4**, 377–385

HARPER, J. L. (1956). The evolution of weeds in relation to resistance to herbicides, *Proceedings 3rd British Weed Control Conference*, pp. 179–188

HIBBETT, C. J. (1969). Growth and spray retention of wild oat and flax in relation to herbicidal selectivity, *Weed Res.*, **9**, 95–107

HOFFMAN, O. L. (1962). Chemical seed treatments as herbicidal antidotes, *Weeds*, **10**, 322–323

HOLLIDAY, R. J., PUTWAIN, P. D. and DAFNI, A. (1976). The evolution of herbicide resistance in weeds and its implications for the farmer, *Proceedings 1976 British Crop Protection Conference – Weeds*, pp. 937–946

HOLLIDAY, R. J. and PUTWAIN, P. D. (1977). Evolution of resistance to simazine in *Senecio vulgaris*, *Weed Res.*, **17**, 291–296

HOLLOWAY, P. J. (1970). Surface factors affecting the wetting of leaves, *Pestic. Sci.*, **1**, 156–163

HOLLY, K. (1976). Selectivity in relation to formulation and application methods, in *Herbicides* (ed. L. J. Audus), vol. 2, Academic Press, London, p. 564

KING, J. M. and HANDLEY, R. P. (1976). Investigations into the addition of mineral oil to bentazone for post-emergence weed control in dwarf beans, *1976 British Crop Protection Conference – Weeds*, pp. 923–930

LUPTON, F. G. H. and OLIVER, R. H. (1976). The inheritance of metoxuron susceptibility in winter wheat, *Proceedings 1976 British Crop Protection Conference – Weeds*, pp. 473–478

LUTMAN, P. J. W. and DAVIES, E. L. C. P. (1976). Investigations into the effect of variety on the susceptibility of potatoes to metoxuron, *Proceedings 1976 British Crop Protection Conference – Weeds*, pp. 491–498

McCAVISH, W. J. (1980). Herbicides for woody weed control by foliar application, *Proceedings 1980 British Crop Protection Conference – Weeds*, pp. 729–737

MAFF (1979). *Review of the Safety for Use in the U.K. of the Herbicide 2,4,5-T*, Ministry of Agriculture, Fisheries and Food, London, p. 14

MAFF (1980). *Approved Products for Farmers and Growers*, Ministry of Agriculture, Fisheries and Food, London, p. 330

MARTIN, J. T. and JUNIPER, B. E. (1970). *The Cuticle of Plants*, Edward Arnold, London, p. 339

MERRITT, C. R. (1980). The influence of application variables on the biological performance of foliage-applied herbicides, in *Spraying Systems for the 80's* (ed. J. O. Walker), Monograph 24, British Crop Protection Council, pp. 35–43

NAYLOR, A. W. (1976). Herbicide metabolism in plants, in *Herbicides* (ed. L. J. Audus), vol. 1, Academic Press, London, p. 608

PALLOS, F. M. and CASIDA, J. E. (1978). *Chemistry and Action of Herbicide Antidotes*, Academic Press, New York, p. 171

PEACOCK, F. C. (ed.) (1979). *Jealott's Hill: Fifty years of Agricultural Research 1928–1978*, Imperial Chemical Industries, Bracknell, p. 160 (section on paraquat and diquat, pp. 66–86)

SAGGERS, D. T. (1976). The search for new herbicides, in *Herbicides* (ed. L. J. Audus), vol. 2, Academic Press, London, p. 564

SHEETS, T. J. and KAUFMAN, D. D. (1970). Degradation and effects of herbicides in soils, in *Proceedings FAO International Conference on Weed Control*, Weed Science Society of America, 668 pp.

SMITH, A. E. and SECOY, D. M. (1976). Early chemical control of weeds in Europe, *Weed Sci.*, **24**, 594–597

SOUZA MACHADO, V., BANDEEN, J. D., STEPHENSON, G. R. and JENSEN, K. I. N. (1977). Differential atrazine interference with the Hill reaction of isolated chloroplasts from *Chenopodium album* biotypes, *Weed Res.*, **17**, 407–413

TOTTMAN, D. R. and DAVIES, E. L. P. (1978). The effect of herbicides on the root system of wheat plants, *Ann. appl. Biol.*, **90**, 93–99

TURNER, D. J. and LOADER, M. P. C. (1974). Studies with solubilized herbicide formulations, *Proceedings 12th British Weed Control Conference*, pp. 177–184

TURNER, D. J. and LOADER, M. P. C. (1975). Further studies with additives: effects of phosphate esters and ammonium salts on the activity of leaf-applied herbicides, *Pestic. Sci.*, **6**, 1–10

TURNER, D. J. and LOADER, M. P. C. (1980). Effect of ammonium phosphate and other additives upon the phytotoxicity of glyphosate to *Agropyron repens*, *Weed Res.*, **20**, 139–146

WAIN, R. L. (1954). Selective weed control – some recent developments at Wye, *Proceedings 2nd British Weed Control Conference*, 311–317

WALKER, A. (1977). Herbicide persistence – the weather and the soil, *A.D.A.S. Q. Rev.*, pp. 168–179

WALKER, A., FARRANT, D. M., BRYANT, J. H. and BROWN, P. A. (1976). The efficiency of herbicide incorporation into soil with different implements, *Weed Res.*, **16**, 391–397

WARWICK, D. D. (1976). Factors contributing to the improved simazine resistance observed in oilseed rape, *Proceedings 1976 British Crop Protection Conference – Weeds*, pp. 479–484

## FURTHER READING

AUDUS, L. J. (1976). *Herbicides: Physiology, Biochemistry, Ecology*, vols. 1 and 2, Academic Press, London, pp. 608 and 564

HANCE, R. J. (ed.) (1980). *Interactions Between Herbicides and the Soil*, Academic Press, London, p. 349

HILL, I. R. and WRIGHT, S. J. L. (1978). *Pesticide Microbiology*, Academic Press, London, p. 844

KIRBY, Celia (1980). *The Hormone Weedkillers*, British Crop Protection Council, London, p. 55

MILBURN, J. A. (1979). Water Flow in Plants, Longman, London, p. 225

WORTHING, C. R. (1979). *Pesticide Manual*, British Crop Protection Council, London, p. 655

# 7 HERBICIDE APPLICATION

## 7.1 REACHING THE TARGET

Compounds that are promising under the controlled conditions of laboratory testing may be disappointing in the field because insufficient of the applied active ingredient reaches the ultimate biological target. Herbicide targets are the soil and weed foliage. Leaves often repel spray droplets leading to bouncing or running off and herbicide that reaches the soil may be lost by evaporation, photodecomposition, adsorption or breakdown before being taken up by weed roots (Chapter 6). Losses also occur by detoxification within the plant so that ultimately a small fraction only of the original dose is physiologically active in killing weeds. This loss of expensive chemical between sprayer and site of action is common to other pesticides and an aim of current research is to reduce the consequent wastage, environmental contamination and damage to non-target organisms that results. The initial choice of herbicide and formulation, mode and timing of application, prevailing weather and other factors, all may influence the proportion of herbicide reaching the target and the effect it has when it gets there. For example, quite small differences in the droplet size spectrum, concentration of wetting agent or other adjuvant, or in air temperature or relative humidity before, during or after spraying, can all have profound effects on the retention of herbicide on leaves and subsequent uptake, movement and behaviour within the plant (Chapter 6).

With so many variables it follows that manufacturers' recommendations are a compromise aimed at achieving good weed control while avoiding crop damage or spray drift, but inevitably much spraying is done under suboptimal conditions. For example, the tolerance of cereal crops to growth regulator herbicides depends on the state of development of the ears, but study has shown that differences in stage of ear development between plants and between tillers on single plants are considerable, reducing further the safe period for herbicide spraying (Tottman 1976; Tottman and Makepeace, 1979).

During the period between the discovery of the growth regulator weedkillers in the 1940s and the provision of simple spraying machines, dust formulations were frequently used. Dusts, which have particles up to about 100μm, give unreliable results and often cause problems due to drift. More recently solid formulations of some soil-applied herbicides with particles generally larger than 300 μm, known as granules, have proved to be advantageous for some situations.

## 7.2 SPRAYING

### 7.2.1 The spraying machine and its origins

When fungicides and other pesticides were discovered in the nineteenth century a simple means of applying them to plants had to be devised. Initially, brooms made from bundles of twigs were first dipped in the liquid – generally Bordeaux Mixture at this time – then shaken vigorously over the target. Later, more elaborate brooms fed by liquid under pressure were developed and towards the end of the century a system of pumps and nozzles was devised. Despite the many refinements and improvements made since them, the system in use today is essentially the same.

Spraying, in which the pesticide reaches the target in a water carrier, remains the most generally used method of application everywhere. The active ingredient is rarely in a form suitable for direct use (Chapter 6) so it must be formulated by the manufacturer into a product that will mix easily with water in the spray tank to form a solution, suspension or emulsion. The liquid is then forced through nozzles under pressure to make it break up into drops. Generally the drops are carried to the target by a combination of hydraulic pressure and gravitational forces, but an air stream created by fans may also be used to carry the drops to the target, although this is rarely done with herbicides. Most modern pesticides are more demanding of the application technique than were earlier ones, and advisers often find farmers and growers applying costly pesticides through unsuitable or poorly maintained spraying machines. Application is the last link in the chain of events from discovery of a new herbicide to its practical use in the field and often it is the weakest link of all. *Figure 7.1* shows some of the factors that contribute to this weakness.

Spraying equipment now comes in a great variety of sizes and types from small hand-held and knapsack machines (MAFF, 1979) to large tractor-drawn ones capable of spraying many hectares each hour (*Figures 7.2 and 7.3*). Sprays are applied from helicopters and aircraft, and use is sometimes made of overhead irrigation as a ready made spraying system for applying pesticides. There are also specialised sprayers such as the 'Arbogard' which delivers a repeatable volume and pattern of herbicide around the base of young trees or shrubs while keeping the spray away from tree foliage or bark (*Figure 7.4*).

Hydraulic pressure can be produced by several types of pump, the pressurised liquid being forced through a nozzle designed to produce

| *Factors that influence the success of a herbicide application* | *Operation* | *Herbicide losses between sprayer and target* |
|---|---|---|
| Design and maintenance of spraying machine.<br>Training of machine operator.<br>Choice of herbicide, formulation or method of application. | Herbicide in sprayer tank.<br>↓<br>Formation of spray droplets.<br>↓ | |
| Weather factors including wind, rain and relative humidity. | Movement to the target, due to hydraulic pressure, gravity, moving air stream or electrostatic attraction.<br>↓ | Drift of small droplets; droplets that miss the target. |
| Formulation, volume rate and droplet size.<br>Leaf surface features.<br>Leaf orientation.<br>Rain or dew. | Retention on leaf surface.<br>Redistribution on plant surface.<br>↓ | Large droplets bounce and splash.<br>Run-off.<br>Volatilisation.<br>Photodegradation. |
| Formulation.<br>Rain, dew.<br>Light, dark.<br>Growth phase of plant. | Uptake and distribution within plant.<br>↓<br>Ultimate biological target within weed plant. | Breakdown or immobilisation within plant. |

**Figure 7.1** Some of the factors that influence the proportion of foliar-applied herbicide that reaches the ultimate target.

either a hollow cone of spray or a flat fan; both cone and fan nozzles are equally satisfactory provided they are of good quality. If several nozzles are mounted together on a boom they must all be of the same specification, be designed for overlapping and be held at the distance above the target specified by the manufacturer. When a narrow strip only is to be sprayed then a nozzle designed to give an even output across the full width of the spray is necessary; larger strips, such as those in which plantation crops are grown, can be sprayed from one side with an off centred nozzle; local application to patches of weed requires a narrow angled solid cone nozzle for best results. Because spray liquids can erode nozzle orifices and thereby increase and distort the nozzle output, they should be checked periodically and if necessary replaced. The construction and operation of all types of sprayers is described by Matthews (1979).

Recommendations for herbicide use are given in terms of amount of herbicide per unit area of ground; those in official manuals will be in kg/ha of active ingredient, while those given by manufacturers refer to

**Figure 7.2** Simple, robust knapsack sprayer of a type widely used for applying herbicides in crop and amenity situations. (By courtesy of Cooper, Pegler and Co. Ltd.)

**Figure 7.3** A modern field crop sprayer with a 550 l tank and a folding 10 m boom designed to spray up to 720 l/ha. (By courtesy of E. Allman and Co. Ltd.)

**Figure 7.4** The 'Arbogard', a sprayer designed to apply herbicide to the weeds and soil around the base of trees and shrubs; the guard prevents spray reaching leaves or green bark of the plant. In the photograph part of a shrub can be seen above the top of the guard. (By courtesy of E. Allman and Co. Ltd.)

weight or volume of the formulated product per unit ground area. This quantity of herbicide will be diluted so that the spray is applied in a given volume per unit area, usually between 600 and 200 l/ha for conventional hydraulic sprayers but volume rates as low as one or two l/ha can be applied from aircraft, or some kinds of spinning disc ground sprayers (see below).

### 7.2.2 The significance of droplet size and coverage

Conventional nozzles produce a spectrum of drop sizes, from those of one or two microns to larger ones several hundred microns in diameter. In order to compare the output of different nozzles some means of defining the size spectrum is required. Several methods of determining the size of individual drops in a spray can be used, ranging from simple embedding of drops in an oily medium and direct measurement under a microscope to electronic devices employing lasers (Matthews, 1979). When the size range has been found the full spectrum can be specified for any particular nozzle as the median diameter, defined as the diameter of the droplet that divides the spray into equal parts, half being in smaller and half in larger droplets than the median. A median can be specified on the basis of either numbers or volume of drops, known as the Number Median Diameter (NMD) or the Volume Median Diameter (VMD), respectively. However, the VMD provides only an approximate indication of the drop size spectrum because a small increase in the proportion of very large drops will cause a disproportionate increase in the VMD and give a distorted view of the size distribution of the spray. Equally, the NMD can give an unsatisfactory indication of the size of the total spray if there are many very small drops present. Matthews (1979) suggests that a better measure of the uniformity of size of the drops making up the spray is given by the ratio between the VMD and the NMD; the nearer the ratio is to 1.0 the more uniform the size of the drops.

The biological effectiveness of a spray depends on many factors but two important ones are the spectrum of droplet size and the degree of coverage with spray of the target that is achieved. At very high volume rates of application individual drops coalesce and excess spray rolls off the leaves. When lower volume rates are applied there is an interrelationship between the drop size and the coverage, such that at any given volume rate per unit area, smaller drops provide superior coverage (*Table 7.1*). But large drops bounce or break up on impact and some of the small drops thereby created are reflected from the plant surface. At the other end of the size spectrum the smaller drops have such a low terminal velocity that they drift away from the target (*Table 7.2*), so that in practice a compromise is sought that combines maximum retention with minimum drift. Although drops below 100 μm may form a small part only of the total spray volume, their

*Table 7.1*

*Theoretical density of droplets between 10µm and 1000µm when uniformly applied to a flat surface at 1 l/ha. (Adapted from Matthews, 1979.)*

| *Diameter of droplets in µm* | *Approximate number of droplets/cm²* |
|---|---|
| 10 | 19 000 |
| 20 | 2 350 |
| 50 | 150 |
| 100 | 20 |
| 200 | 2.5 |
| 400 | 0.3 |
| 1000 | 0.02 |

*Table 7.2*

*Terminal velocity and fall time in still air of spray droplets of specific gravity 1.0 dropped from 3 m. (From Matthews, 1979.)*

| *Diameter of droplets in µm* | *Terminal velocity in m/s* | *Fall time (approx.)* |
|---|---|---|
| 10 | 0.003 | 17 min |
| 20 | 0.012 | 4 min |
| 50 | 0.075 | 40 s |
| 100 | 0.279 | 11 s |
| 200 | 0.721 | 4 s |
| 500 | 2.139 | 1.5 s |

relatively slow fall and consequent drift downwind can cause severe damage to susceptible crops nearby. For example, tomatoes in glasshouses have been damaged by herbicides applied several kilometres away; slight damage due to such drift is seldom diagnosed correctly, but it is known that appreciable yield reductions in susceptible crops can be caused.

An ideal droplet size for each operation cannot yet be specified, and the relatively wide range of drop size produced by nozzles gives conventional hydraulic sprayers a flexibility not shown by some of the more recent spraying devices, such as spinning disc equipment, which produce drops of extreme size uniformity (Figure 7.5).

### 7.2.3 Droplet formation in sprayers

Nozzles emit flat or cylindrical sheets of water which first break up into ligaments; the same physical forces that create ligaments then cause drops, separated by smaller satellite droplets, to form from them. An inevitable consequence of this mode of drop formation is that a broad spectrum of drop size is present in the spray.

Other methods of drop formation have been devised to avoid the formation of small drifting droplets. One method is to employ vibrating plates or tubes pierced by small holes through which spray liquid is forced under low pressure. These devices largely eliminate the very small droplet fraction, and although they have the disadvantage that many of the drops produced are too large for good retention on the target, they have been successfully used for selective paraquat application where all contact with the crop must be avoided. Similar results have been achieved with anvil or deflector nozzles working at less than 0.7 bar, but these nozzles deliver more spray at the edges than at the centre of their spraying width and therefore they are less suitable for situations in which the dose rate is critical.

### 7.2.4 Controlled droplet size spraying

Using an entirely different method of drop formation, Bals (1975) has succeeded in reducing the variability in droplet size. Spray liquid which is fed on to a radially grooved spinning disc or cup is thrown off the finely toothed edge as remarkably uniformly sized drops (*Figure 7.5*). By using discs of differing form and varying the speed of rotation, drops from about 50 μm for killing flying insects to 200–350 μm for herbicides can be produced. Machines such as the Micron 'Herbi' (*Figure 7.5*) can apply herbicides in drops exceeding 200 μm, eliminating drift altogether, and because all the drops are of an appropriate size for the target the volume rate can be reduced from about 200 l/ha to about 20 l/ha; the term Controlled Droplet Application (CDA) has been coined for such spraying. Insecticide spraying with spinning discs producing drifting droplets has enabled substantially reduced rates of active ingredient to be used but similar reductions have not so far been possible with CDA spraying of larger, non-drifting drops. When lower volume rates are used the spray solution is correspondingly more concentrated, and to avoid ingestion or inhalation the drops should not be smaller than 200–250 μm. CDA application has advantages where the volume of spray that can be carried is critical or where fresh clean water is unavailable, but it seems unlikely to replace established methods for field spraying in the immediate future (Chapters 11 and 12) (Linke, 1978).

### 7.2.5 Application through irrigation equipment

Soil-acting herbicides such as EPTC and some triazines can be applied at extremely high volume by injecting herbicide into the water used for overhead irrigation. Although droplet size is relatively large, the greater height reached renders droplets susceptible to drift and uniform application cannot be expected when wind speeds exceed

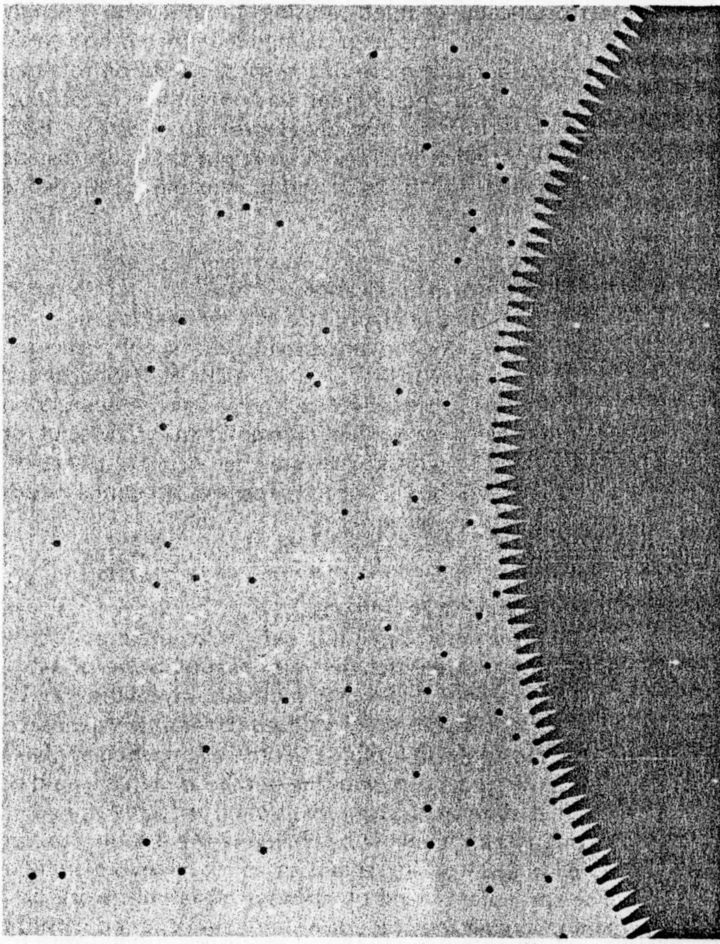

**Figure 7.5** Droplets of 250 µm (greatly magnified) leaving a spinning 'Herbi' disc. Note that they form individually on the teeth of the disc and that they are uniform in size. The disc is spinning at 2000 rev/min. (By courtesy of Micron Sprayers Ltd.)

about 15 km/h. After EPTC has been applied by overhead irrigation, sufficient water can then be sprayed to seal the herbicide in the soil.

### 7.2.6 Electrostatic charging of spray droplets

For herbicides applied to foliage it has been suggested that the drops should be electrostatically charged causing them to be strongly attracted to the foliage, and it has been claimed that the method could improve the coverage because more of the spray would be retained on the target. The necessary equipment was potentially hazardous, bulky and expensive, but recent developments (Coffee, 1980) (Chapter 12) using light electronic devices suggest that electrostatic sprayers may eventually be both cheap and practicable for herbicide application.

### 7.2.7 Other novel methods of application

Non-selective herbicides such as glyphosate may be used in a selective fashion if the active ingredient contained in liquid or a gel is smeared rather than sprayed on to weeds that are taller than the crop. Several types of machine with wet ropes or wet spongy rollers are being tried (Chapter 12). None are likely to supplant conventional sprayers but for application to unwanted species in turf growing above a predetermined height, to sugar beet bolters or to patches of perennial weeds within a crop, they have many advantages.

### 7.2.8 Calibration of spraying equipment

All sprayers require calibration since it is only by accurate control of the volume rate per unit area that the rate of application of active ingredient can be guaranteed. Several methods of calibration for different types of sprayer have been described (Fryer and Makepiece, 1977; MAFF, 1979); most rely on finding the time taken when applying a known volume of the spray, then working out the area that would be covered during that time at a given controlled working speed. If necessary, adjustments can be made by changing the nozzle size, the spraying pressure or the working speed so as to achieve a suitable volume rate. The spray trains used on railways sometimes work on a different principle, varying the amount of herbicide applied with the speed of the train, so as to spray the same rate of herbicide per length of track irrespective of forward speed.

The small or irregularly shaped areas common in amenity plantings create problems in accurate application (Chapter 11) and it may be necessary to mark the sprayed area in some way to prevent overdosing.

## 7.3 GRANULES AND THEIR APPLICATION

Herbicide granules contain the active ingredient diluted ten or more times with a relatively inert solid carrier such as clay. Most granules are made either by spraying the toxicant on to preformed pellets or by moulding from a dough-like mixture of carrier and toxicant. The second or agglomeration method allows a controlled rate of release of toxicant to be built in to the granule by appropriate choice of carrier and additives, also agglomeration granules can be made which contain a higher concentration of active ingredient so making them cheaper to manufacture, transport and store. Different kinds of granules vary in size between 250μm and 3,700μm but each kind should be as uniform in size as possible for efficient application. A dust fraction is produced by crushing or abrasion and not only may this contain a higher concentration of active ingredient but it will be liable to drift downwind.

The main use of herbicide granules has been for the slow release of total herbicides in industrial sites, but their use in amenity horticulture for aquatic weed control and in crops is increasing. Apart from the obvious advantage of requiring no water for dilution, granules of highly toxic pesticides (so far mainly insecticides) are less hazardous to apply than sprays, granules produce fewer drift problems, and their ability to penetrate a canopy of foliage enables a herbicide applied overall to reach the soil. However, the distribution achieved with granule applicators is generally less accurate and uniform than with a good sprayer.

The release of active ingredient from the granule by leaching or volatilisation will depend on the physical properties of the herbicide used and on the way the granule is made, and ideally the toxicant should be released at a rate that ensures an adequate concentration in the root zone of the weed (Furmidge *et al.*, 1966).

However, this may be difficult to achieve because the herbicide, once released from the granule, will be subject to adsorption, decomposition, volatilisation and leaching in the same way as will a herbicide that has been applied by spraying. Small size, which improves coverage and reasonably uniform distribution on the soil, will be most important for herbicides which become strongly adsorbed on soil colloids. According to Furmidge (1972) the granule size may be more important in achieving biological effectiveness than either the percentage of active ingredient or the dose rate for such herbicides. By contrast, where the herbicide is not strongly adsorbed some lateral movement is possible and particle size and uniformity of distribution become less important.

As granules can be made in several ways from many different materials, their density and other physical properties can be equally variable. Therefore it is essential to recalibrate machinery each time granules are used. Small hand-held applicators relying on gravity feed are ideal for spot treatment, for example, applying dichlobenil granules to localised patches of herbaceous perennial weeds. There are also hand-driven or motorised knapsack distributors suitable for applying a continuous band of granules beneath trees or for total weed control. Larger machines, for agricultural and industrial use, have the granules metered from a hopper at a rate proportional to forward speed and then carried in an air stream to the distributing nozzle.

## 7.4 GOVERNMENT REGULATIONS AND APPLICATION

Although rather few herbicides possess acute oral or dermal toxicity to man, reasonable care should always be taken to avoid contamination of skin and clothing. In many countries protective clothing must be worn when applying the more toxic pesticides, with minimum standards set out in official regulations. In Britain, for instance, the Health and Safety (Agriculture) (Poisonous Substances) Regulations 1975 require users of certain listed toxic pesticides to observe a basic code of good practice (MAFF, 1980). Most such toxic materials are insecticides and the only listed herbicides requiring elaborate protective clothing are DNOC and dinoseb. The code of practice has provisions for safe storage, the keeping of detailed records when toxic materials are used, the supply of protective clothing and adequate washing facilities. Inspectors are required to enforce these regulations (Anon, 1978).

Large areas are now sprayed with pesticides from the air and inevitably some spray reaches sites for which it was not intended, but British contractors now work under the strict supervision of the Civil Aviation Authority. In Britain the only herbicides approved for application from the air in 1980 were asulam for controlling bracken in grassland and chlortoluron, methabenzthiazuron and terburyne, all for grass weeds in cereals; the application of several fungicides and insecticides was also approved (MAFF, 1980).

## REFERENCES

ANON (1978). *Poisonous Chemicals on the Farm*, HMSO, London, p. 38

BALS, E. J. (1975). Development of a CDA herbicide handsprayer, *PANS*, **21**, 345–349

COFFEE, R. A. (1980). Electrodynamic spraying, *Spraying Systems for the 1980's*, Monograph 24, British Crop Protection Council, London, p. 317

FRYER, J. D. and MAKEPEACE, R. J. (eds.) (1977). *Weed Control Handbook*, Vol. 1, *Principles*, Blackwell, Oxford

FURMIDGE, C. G. L. (1972). General principles governing the behaviour of granular formulations, *Pestic. Sci.*, **3**, 745–751

FURMIDGE, C. G. L., HILL, A. C. and OSGERBY, J. M. (1966). Physicochemical aspects of the availability of pesticides in the soil. 1. Leaching of pesticides from granular formulations, *J. Sci. Fd. Agric.*, **17**, 518–525

LINKE, W. (1978). CDA – a review of developments to date, *Proceedings 1978 British Crop Protection Conference – Weeds*, pp. 1047–1057

MAFF (1979). *Horticultural Sprayers for Small Areas*, booklet 2070, Ministry of Agriculture, Fisheries and Food, London, p. 21

MAFF (1980). *Approved Products for Farmers and Growers*, Ministry of Agriculture, Fisheries and Food, London, p. 330

MATTHEWS, G. A. (1979). *Pesticide Application Methods*, Longman, London, p. 334

TOTTMAN, D. R. (1976). Spray timing and the identification of cereal growth stages, *Proceedings 1976 British Crop Protection Conference – Weeds*, pp. 791–800

TOTTMAN, D. R. and MAKEPEACE, R. J. (1979). An explanation of the decimal code for the growth stages of cereals with illustrations, *Ann. appl. Biol.*, **93**, 221–234

## FURTHER READING

GROENWOLD, B. E., PEREIRO, F., PURNELL, T. J. and SCHER, H. B. (1980). Microencapsulated thiocarbamate herbicides: A review of their physical, chemical and biological properties, *Proceedings 1980 British Crop Protection Conference – Weeds*, pp. 185–191

OUTLOOK ON AGRICULTURE, **10** (7) (1981). This issue is devoted entirely to spray application technology.

PROCEEDINGS OF *Symposium on Controlled Drop Application* (1978). Monograph 22, British Crop Protection Council, London, p. 275

WALKER, J. O. (1980). *Spraying Systems for the 1980's*, Monograph 24, British Crop Protection Council, London, p. 317

# 8 HERBICIDES AND CROP AND SOIL MANAGEMENT

## 8.1 ECONOMIC AND SOCIAL CONSEQUENCES OF HERBICIDE USE

Effective herbicides now enable weeds to be controlled without cultivation. Despite their high development costs the flow of new herbicides continues and there are few crops in which chemical weed control is not possible. The drudgery of hand weeding has been virtually eliminated in the technically advanced countries but the herbicides used in less highly developed countries are almost exclusively applied to plantation cash crops; peasant agriculture relies still on hand weeding although great social benefit could be derived from carefully planned use of quite small quantities of herbicides, such as the use of triazines for suppression of parasitic weeds in maize (Ogborn, 1972).

Weed control with herbicides is usually much cheaper than using hand labour and this has been the main incentive for their widespread adoption. A comparison made in Northern America, typical of results everywhere with labour intensive crops, showed that the cost of weed control by mechanical cultivation and hand weeding of one hectare during the lifetime of one crop was almost eight times the cost of chemical weed control during the same period (Dowler and Hauser, 1975). If herbicides were no longer available a large increase in food and fibre prices would be inevitable.

By reducing hand work, herbicides have led to fewer people being engaged in food production, and in the United States, for example, little more than two per cent of the population produce the nation's food. Crops that formerly were labour intensive, including peas, carrots and strawberries, are now grown extensively as farm crops; tree and bush fruit crops are being managed without cultivation, using novel methods; farm crops can now be drilled and grown to maturity without cultivation, and some crops that previously were transplanted from a seedbed can now be sown direct. Herbicides have been responsible for revolutionary changes in soil management and research in many countries is now monitoring the long-term effects of the changes that are occurring (van der Zweep, 1971).

## 8.2 MANAGEMENT CHANGES FOLLOWING WEED CONTROL WITH HERBICIDES

### 8.2.1 Need for crop rotation

The value of leguminous crops in fixing nitrogen for the benefit of succeeding crops has long been known in principle, but the notion that regular crop rotation within a mixed farming system is also desirable, is a comparatively recent one. Paddy rice has been grown continuously in the same fields for thousands of years in Southeast Asia, but here the build up of weeds is prevented by transplanting from a seedbed into flooded land; fertility is maintained by returning all organic wastes to the rice fields. With dry land crops the increase in weed populations has always been a major constraint preventing continuous cereal production and the introduction of a turnip 'break crop' instead of a non-productive fallow enabled these weed populations to be contained without loss of yield; the notions of 'fouling' and 'cleaning' crops, now almost forgotten in Britain, were well understood by farmers less than a generation ago. The adoption of routine herbicide application to grain crops preceded their introduction for break crops and at first the position was reversed, making cereals into a cleaning crop; but now with the discovery of new selective herbicides for root crops there is no longer a necessity to suffer from fouling crops. Grass weeds inevitably increased in cereal crops following the suppression of broad-leaved weeds by hormone herbicides but fortunately it is now possible to control Black-grass (*Alopecurus myosuroides*), Wild-oats (*Avena* spp.) and Common Couch (*Agropyron repens*); in Britain the increase in Barren (or Sterile) Brome (*Bromus sterilis*) in cereals remains an unresolved problem. In arable horticultural crops regular rotation is rarely practised, but repeated plantings of lettuce or brassicas that are treated with selective herbicides does cause a build up of weeds botanically related to these crops, such as Mayweeds in the former and cruciferous weeds in the latter. An equally important reason for avoiding the continuous growing of any single crop is the aggravation of pest and disease problems that often follows, and which in the absence of weed difficulties may be crucial in the decision to introduce a break crop. Legislation already exists in Britain and elsewhere to ensure that fields infested with certain eelworms should not be planted with susceptible crops until the eelworm population has fallen to an acceptable level.

### 8.2.2 Crop spacing

This is determined in part by the need for weed control because of the space required by men, draught animals or tractors while controlling weeds. Bleasdale (1963) described how, when labour was cheaper than it is today, vegetables were broadcast in beds about 1.5 m wide, then singled, thinned and weeded entirely by hand. Although this was possible for high-value, small-scale crops, a less expensive system was required for extensive farm crops, and in the 18th Century Jethro Tull publicised his horse hoeing husbandry in which root crops were sown with drills in straight lines or rows sufficiently far apart for horse hoeing of the inter-row; because only the weeds in the row then remained to be removed by hand there was a big saving in human labour compared with overall hand weeding, and row-cropping quickly became the standard practice, not only for farm crops but for horticultural ones also.

But effective chemical control of weeds has allowed a reappraisal of spacing requirements, often leading to substantial changes in cultural methods. By reducing the distance between pea rows, for example, a change from the former practice of picking several times by hand as the pods matured has given way to a once-over mechanical harvest of fresh peas for processing; crowded small plants each produce fewer pods, but a greater proportion of pods have reached a usable stage at harvest. It was found (Bleasdale, 1963) that the nearer the planting was to being 'on the square', the higher was the yield of peas at an acceptable stage of development; the highest yield was obtained when the plants were at a density of about 65 plants/m$^2$ in rows 0.2 m apart. Similar conclusions have been reached with respect to many other crops, but sometimes compensating factors appear to reduce the advantages of 'on the square' arrangements. Maincrop potatoes, for example, will tolerate considerable differences in the squareness of planting, with row widths up to 1.0 m coupled with close planting within the row giving acceptable yields in North America. When the size of the individual plant is critical as, for example, in lettuce, cabbage or celery, there is little room for manoeuvre in spatial arrangement without sacrificing crop marketability. The control over plant size offered by varying the spacing has, however, been attempted with some success in carrots for processing and small cauliflowers and potatoes for freezing and canning, respectively. Close spacing between the rows provides an added bonus in that the crop, by its rapid completion of a leaf canopy, competes earlier and more effectively with any weeds that survive; improvements in the evenness and speed of establishment of the seedling stand are now being attempted by the use of vigour-tested seed, fluid drilling, seed priming and more accurate control of drilling depth.

In herbaceous perennials and bush and tree crops, too, weed control with herbicides has enabled a reappraisal of spatial arrangement to be made. Closer planting and mechanical harvesting of blackcurrants have required novel cultural systems, and in apples there has been a gradual evolution from widely spaced trees in a cultivated orchard to smaller more densely planted ones that come rapidly into bearing. This is taken to an extreme in the so-called meadow orchard, an experimental system in which the trees are closely spaced and almost reduced to the size of brussels sprouts plants, and all weed and pest control is by spraying from above (see Chapter 10).

### 8.2.3 Reduced cultivation techniques – annual crops

The faith in the benefits bestowed by repeated and thorough cultivation and a general belief that 'crops will suffer and yields drop by just the extent the soil tilth falls short of perfection' was being questioned more than 100 years ago, but with rising costs the questions became more urgent (Keen and Russell, 1937; Bullen, 1977; Cannell, 1977). As recently as 1870, seedbed preparation involved more than one ploughing followed sometimes by as many as 40 other operations, but after investigating the influence of soil cultivation on crop yield, Keen and Russell (1937) concluded that cultivation should be limited to the minimum required to produce a seedbed and for weed control and that further cultivation actually reduced yields. Since then, their findings have often been confirmed.

In North America the impetus to reduce tillage was provided by the harmful effect of cultivation on soil stability – and consequent wind and water erosion. Several techniques were devised for reducing cultivation and keeping plant debris from the previous crop on the surface thereby reducing erosion, but all foundered on the need for ploughing to bury weeds before sowing the next crop. However, herbicides such as paraquat and glyphosate have made non-cultivation systems practicable, and an increasing area of land in many countries is now sown with little or no prior cultivation. In North America crops such as soya beans, cotton, ground nuts, wheat, tobacco and some vegetables, but above all maize, are successfully grown with reduced (minimum) or 'zero' tillage; according to United States Department of Agriculture estimates, by the end of this century over half the crops there will be sown by direct-drilling, and most of the rest will receive much reduced cultivation (Triplett and Van Doren, 1977). Reduced cultivation allows one crop to follow so quickly after the harvest of the previous one that two crops per year can sometimes be grown (Barker and Wünsche, 1977); not only is time saved but the soil retains sufficient moisture to allow the next crop to establish more readily than when it has been repeatedly exposed to the air during cultivation. (Lal *et al.*, 1978.)

In Britain direct-drilling has been successful with farm crops (MAFF, 1978) including wheat, barley, kale and other brassica crops. The yields of direct-drilled cereals have often been less than those of conventionally managed crops for the first year or two but when more cereal crops are grown on the same land these differences have often changed in favour of direct drilling. Although some of the problems with mineral nutrition, pests, diseases and soil compaction that had been predicted have actually occurred, none has been sufficient to prevent a continued increase in the area of direct-drilled crops (Allen, 1975). Perennial weeds, especially rhizomatous grasses, thrive without cultivation, but these can now be controlled with glyphosate. Slugs, too, find the conditions following direct-drilling favourable and poisoned baits may be necessary to prevent them from eating the germinating seeds. Initially, too, there were problems of drill design to be overcome before satisfactory drilling into hard, untilled soil covered with plant debris could be achieved. Another common difficulty with direct-drilled crops is inadequate contact between the germinating seeds and soil particles but this can be overcome by rolling immediately after drilling.

The previous assumption that cultivation was essential, not only for weed control but for soil aeration, to bury manures and plant debris and to prepare a seedbed with a fine tilth, seems not to be fully justified. Direct-drilling is now widely practised in many countries for extensive farm crops (Haywood *et al.*, 1980), but there is as yet little direct-drilling of high-value vegetable crops, and root crops such as parsnips, carrots and especially potatoes seem only to do well after thorough seedbed preparation; also, on soils with a high silt content that are prone to settle with inadequate air space, regular cultivations will probably continue to be required for arable crops (Cannell *et al.*, 1978).

### 8.2.4 Reduced cultivation techniques – perennial crops

Elimination of cultivation is not new in commercial fruit growing, especially in citrus and other plantation crops (Robinson, 1963). A change from cultivation to chemical weed control was begun in Californian citrus orchards to reduce production costs, but in addition the fruit produced has often been of better quality, has matured earlier and yields have been higher. Initially, weed control was achieved with petroleum oils, which were rather more expensive than traditional methods until the reduced numbers of weed seedlings that emerged in the absence of cultivation enabled lower annual rates of oils to be applied; more recently the costs have been further reduced by using persistent soil-acting triazine and substituted urea herbicides.

Chemical weed control in temperate fruit crops was not attempted until soil-acting herbicides became available in the late 1950s. But because the costs of weed control in crops such as strawberries, blackcurrants, raspberries, grapevines and asparagus were already high, growers were eager to adopt the new materials, although at first their use was poorly integrated into the whole management system. Simazine, for example, kills germinating seedlings but has less effect on established plants, and its use in poorly managed plantations frequently led to their becoming overrun with perennial weeds. However, this situation is now easily avoided by using spot treatments of dichlobenil, chlorthiamid, oxadiazon or glyphosate. At first, too, growers failed to appreciate that effective long-term weed control meant that cultivations were no longer required, and the present non-cultivation systems were only gradually evolved.

In apples the general practice of frequent cultivation between the trees gave way to a permanent and closely mown grass sward, the trees being set in squares or strips that were kept weed free with herbicides. There is now an increasing interest in overall chemical weed control, particularly with the small, closely spaced trees now popular because they come into bearing more rapidly and are easier to pick, prune and spray. Bare soil loses less water than a sward, but a surface mulch is necessary, particularly on sloping land, when the site is prone to erosion. Growers still prefer to have grass strips between the tree rows, but when herbicides are applied overall there is more rapid tree growth and higher yields are usually obtained (Chapter 10).

The greatest advantage from herbicides has been gained in crops such as strawberry in which hand weeding is laborious and expensive. In this crop, all operations except picking have now been mechanised, making it ideal for farmers growing for the pick-your-own trade.

### 8.2.5 Direct sowing instead of transplanting

Transplanting from a seedbed to permanent quarters is more time consuming than drilling followed by thinning and singling, but its main advantage lies in the ease and cheapness of weed control on the large uncropped area; expensive hand work is only necessary in the seedbed. A modified method of great antiquity is that used in paddy rice production, the field being flooded to control weeds while the rice seedlings are grown nearby in a seedbed. Rice, which is derived from marsh grasses and is able to grow well when planted in water, can then be transplanted into the soil through a weed-controlling layer of water.

Planting of frost-sensitive species, such as tomato and sweet corn, which have been raised under frost protection, enables larger outdoor crops to be obtained than would have been possible by sowing in the open ground after the danger of frosts has receded. Another advantage of transplanting lies in the ability to set out uniformly sized plants at the required spacing thus ensuring an even stand. Furthermore, an operation that was labour demanding is now less so since the advent of transplanting machines.

The principal disadvantage of transplanting, apart from generally greater costs, is that plants lifted from a seedbed with few roots receive a shock, especially when planted into dry soil, and in the cauliflower, for example, the stress of transplanting can induce premature development of small and useless heads. Interest has also been generated in the direct sowing of crops such as leeks and brussels sprouts into their final stations, maintaining freedom from weed competition with herbicides. Most direct-sown crops will require the sequential use of several herbicides or herbicide mixtures, at least until the crop is itself able to offer some competition to further growth of weeds. Precision drilling is essential to provide a satisfactory stand of plants without the use of hand labour for thinning, but some gaps due to germination failure are inevitable. In an attempt to overcome the problems of plant failure after direct-sowing, and to minimise differences in individual sizes of plants in transplanted crops, new techniques such as planting of small seedlings or pre-germinated seeds in small soil blocks or in gels, respectively, are being developed.

In the rice crop a change from transplanting to direct-sowing is occurring, especially in Japan, where there is a great potential for the saving of labour. Elsewhere the change has been confined to horticultural crops.

### 8.2.6 Parks and other amenity areas

Those responsible for the upkeep of amenity areas (Chapter 11) have been forced by financial stringency and shortage of labour to economise on maintenance costs. Use of herbicides has greatly reduced the high labour charges for weeding on rockeries and among bedded out plants, herbaceous perennials and trees and shrubs. Certain species and cultivars are especially sensitive to particular herbicides, and although there is sufficient data available to enable a choice of herbicide tolerant plants to be made (Fryer and Makepeace, 1978) more information on the relative susceptibilities of different plants is needed.

Regular grass cutting is another large expense in amenity area management. When the cost of frequent cutting is reduced by cutting less often then the cut grass may have to be removed, so that little or no net saving is made. Some local authorities have reported success with maleic hydrazide (Waterhouse, 1980) with savings of almost 50 per cent. Previous experience had suggested that treatment with maleic hydrazide was only suitable for inaccessible and infrequently visited areas, but recent work suggests that it may be possible to combine the retardent effect with attractive visual appearance of the turf (Chapter 11).

Many authorities now save hand labour by spraying weeds that appear between paving stones, penetrate paths and grow beside road verges. The use of herbicides near to private gardens is always open to some risk, and therefore this kind of work should be done only by adequately trained staff using suitable equipment. Appropriate herbicides can now provide control of annual and perennial weeds amongst most kinds of woody plants; herbaceous ornamentals are more sensitive but simazine and lenacil can be used safely in many situations. However, because certain species and cultivars are particularly sensitive to some herbicides, manufacturers should first be consulted before spraying ornamental plants whose tolerance is in doubt. Self-sown seedlings and suckers, which are also resistant to some soil-acting herbicides, can become a secondary weed problem after long periods of herbicide treatment (Chapter 11).

Rose growers, nurserymen, foresters and some local authorities now use herbicides routinely. However, there remains a widespread fear among amenity horticulturists that their plants will suffer if herbicides are used, and this has restricted the more widespread adoption of chemical weed control.

## 8.3 POTENTIAL ENERGY SAVINGS WHEN HERBICIDES ARE USED

The increasing costs and diminishing known reserves of oil and coal have raised questions about the most economical use of the support energy inputs derived from fossil fuel required by advanced and high yielding agriculture. Studies have shown that the support energy consumed in food production in the developed world may greatly exceed its calorific value.

What contribution can herbicides make to energy saving? The on-farm energy inputs are for fertilisers, machinery (manufacture and running costs) and pesticides. Pesticides, including herbicides, are relatively energy demanding in manufacture, but because so little is used the total energy cost per hectare of fertilisers is at least 30 times greater. The main energy costs are from fertilisers (nitrogen fixation being especially energy consuming) and as fuel for tractors (Green and McCulloch, 1976).

Reduced cultivation systems can contribute to a saving in tractor fuel, and estimates made in Britain by Green and McCulloch (1976) show that pre-planting paraquat application and direct-drilling, when compared with conventional cultivations, generally shows a reduction of 50 per cent in expenditure of tractor fuel, after making allowance for the energy equivalent of the herbicide; similar figures have been obtained in North America for 'zero-tillage' of maize. The substitution of herbicides for in-crop mechanical cultivation can make a saving of up to 75 per cent in energy terms, depending on the herbicide used. An indication of the huge size of the total earth moving operation that herbicides could replace is given by an estimate that the annual movement of soil during crop cultivation in the United States is equivalent to a ridge of soil 30 m high, 1.6 km wide, stretching from New York to San Francisco (Alder *et al.*, 1976) (see also Patterson *et al.*, 1980 in Chapter 12).

Although herbicide production is an efficient way of using energy in food production, other large energy costs remain, especially for making nitrogen fertilisers. Also, in developed countries, further large amounts of energy are used in transport, in processing the farm produce into human food and in sophisticated preservation and packaging. Yet the total energy usage for food production on farms in industrial countries is generally less than 5 per cent of the total national usage and therefore savings could probably be made more easily in industry, transport and in domestic heating. The problem is more immediate and acute, however, in countries like India that are dependent upon imported fuel oil and in which agriculture is a major national consumer of energy.

## 8.4 BIOLOGICAL CONSEQUENCES OF HERBICIDE USE

### 8.4.1 Effects on soil structure and fertility

A distinction can be made between herbicides used merely as an alternative way of controlling weeds within a short duration crop, and their use to eliminate, or greatly reduce, soil cultivations over several years. The latter may occur both with plantation crops grown without cultivation and with continuously grown cereals that are direct-drilled into the stubble following herbicide spraying. It seems improbable that substitution of chemical weed control for light cultivations within a crop would produce long-lasting effects on soil structure or fertility (Fryer *et al.*, 1980), although short-term effects can be important. In potatoes, for example, use of herbicides in place of post-planting cultivation can lead to a beneficial reduction in the formation of durable soil clods, of similar size and density to potatoes, that cannot easily be separated from the tubers by mechanical harvesters; and in frost-sensitive crops, including maize and potatoes, untilled bare soil can sometimes give up sufficient extra heat during the night, compared with an equivalent tilled soil, to prevent frost damage. When untilled soils are covered with a layer of organic debris, however, the opposite is found, and sowing of frost-sensitive crops in the spring may have to be delayed.

In nature cultivation is restricted to the activities of animals, especially worms, and this does not lead to loss of soil structure or fertility. On the contrary, soil is given stability by an extensive system of living and dead roots, and freshly ploughed grassland generally possesses abundant organic matter associated with an excellent structure which is gradually reduced under cultivation. Land that has been regularly cultivated for long periods, such as Broadbalk at Rothamsted Experimental Station, reaches an equilibrium at a lowered organic matter content; at this reduced level, with poorer structure, it still produces reasonable crops provided that it is adequately fertilised. However, even in Britain, but especially in some parts of Asia, Africa and the Americas, where soil temperatures are higher and rain can fall with much greater intensity, soil that is frequently cultivated and kept bare of vegetation and plant debris becomes prone to water and wind erosion, making the maintenance of good soil structure essential (Scott Russell, 1977; Triplett and Van Doren, 1977).

Comparisons that have been made between the effects on the soil of different management systems in perennial crops generally show that soil bulk density is decreased under grass and by cultivation and increased under a non-cultivation regime (Atkinson and Herbert, 1979). However, the decrease following cultivation is often short-lived when an unstable soil settles, and the bulk density six months after a cultivation may well be much the same as on an adjacent non-cultivated area. The increased value in bulk density in the upper layers of non-cultivated soils remains much less than the value of about 1.8 g $ml^{-1}$ that would be required to prevent root extension, and it is commonly observed that crop roots ramify freely through this denser soil.

Bulk density alone, however, is an incomplete guide to the soil condition, the distribution and continuity of the pores being more directly related to its ventilation properties. Measurements from soil in fruit plantations indicate that the percentage of large pores greater than 115$\mu$m is equally high under continuous herbicide treatment (non-cultivation) and under grass, and that although soon after cultivation large pores are more numerous, the number rapidly falls to about the same figure as for an equivalent non-cultivated area. To be effective in gas exchange, pores must be part of a continuous system and it seems probable that continuity would be greater in the stable conditions of zero-cultivation than in freshly tilled soil. There is, too, evidence that infiltration rate of rainfall into untilled soil is greater than in corresponding cultivated sites, despite the greater bulk density of the former.

In direct-drilled cereal crops both a reduction in early root growth and a higher soil bulk density often occur. On the other hand, a gradual increase in organic matter near the soil surface when direct-drilling is continued for several years is associated with greater stability of the soil crumbs, which is noticed by the farmer as a superior tilth following drilling. It is probable, too, that channels are left in the soil as old roots die, conferring stability and continuity on the ventilation and drainage properties of soil.

An important argument in favour of ploughing and other pre-planting cultivations has been that they facilitate placement of fertiliser and organic manure into the rooting zone, but results from Britain suggest that even in dry years, roots easily extract nutrients from near the soil surface (Drew and Saker, 1980). In both direct-drilled cereals and in non-cultivated plantation crops, a reduction in soil pH, sometimes sufficient to cause manganese toxicity, can be detected after one or two years and the addition of lime may be necessary. The behaviour of potassium and nitrogen has not been consistent, but phosphorus invariably accumulates in the upper soil layers and levels twice as large as those under cultivated conditions are not uncommon. This unusual distribution appears to be no disadvantage in cereal production, but in plantation crops it may be desirable to incorporate phosphorus in the soil before planting.

### 8.4.2 Effects on soil organisms

Rapid fluctuations in the populations of the many types of soil organism are the norm and it is seldom possible to attribute an observed change to the presence or absence of herbicide residues in the soil. Changes produced experimentally by applying large and unrealistic doses of herbicides may have little relevance to the actual situation.

Fortunately herbicides appear to have few lasting direct effects on soil microorganisms at recommended application rates and such effects as have been detected appear to be indirect ones caused by reduced food supply when weeds are controlled (Greaves, 1979); but because of the importance of microorganisms in soil fertility a watch must be kept for possible harmful effects.

Similarly, the observed population changes in soil invertebrates appear to result largely from changes in the organic food supply (Edwards and Stafford, 1979). Paraquat, a herbicide often applied prior to drilling, appears not to harm earthworms and the destruction of stubble weeds, in addition to the stubble itself, provides abundant plant residues for worms to feed on so that they thrive and multiply; on the other hand, ploughing and cultivation are directly harmful to worms. The substantially larger worm populations existing under direct-drilling regimes may well be of importance in the success of this technique, especially on the heavier soils. However, worm populations will decline if there is a lack of organic debris; in areas such as orchards that are kept continuously weed-free such a decline has been observed, and it may become desirable to add bulky organic matter to maintain worm populations (Atkinson and Herbert, 1979).

A few herbicides, including TCA and DNOC, are directly harmful to worms under laboratory conditions, but field populations are reduced only when excessively high rates are applied. Other soil organisms, such as nematodes and mites, respond in similar ways to the available food supply, and direct effects appear to be relatively slight (Edwards and Stafford, 1979).

### 8.4.3 The effect of herbicide use on the weed flora

The weed flora in arable cropping is subject to outside influences such as the species, cultivar and planting density of the crop, the date of sowing and harvesting, amount of fertiliser and type, depth and frequency of cultivation (Cussans, 1976) as well as the soil type and the climate. By removing the competition from susceptible species, herbicides can cause changes in the species composition of the weed flora, although these changes are generally strongly buffered by the reserves of dormant seeds in the soil. Selection for herbicide susceptibility is possible also within species (Chapter 6) but, because herbicides are used in rotation and again because of the buffering action on selection of unselected dormant seeds, the selection of resistant types is likely to remain of local and temporary importance only.

Changes in on-farm weed populations attributed to the annual application of 2,4-D and MCPA in cereal fields include increases in grass weeds and in Common Chickweed (*Stellaria media*), Knotgrasses (*Polygonum* spp.) and a sharp decrease in Charlock (*Sinapis arvensis*) (Chancellor, 1979). Under experimental conditions in continuous wheat plots, five consecutive annual herbicide applications caused a reduction in the buried weed seeds of susceptible species and an increase in buried seeds of resistant species, such as Scentless Mayweed (*Tripleurospermum maritimum* subsp. *inodorum*) (Thurston, 1969). In a similar experiment at Oxford, however, large differences in the relative abundance of weed seedlings over the period of the experiment were unrelated to herbicide treatments, but were thought to be caused by recent vagaries in the climate in Britain (Chancellor, 1979).

The present evidence suggests tht the weed flora is no more than suppressed during periods without cultivation, and that weed seeds again germinate with much the same species composition as before when cultivation is resumed. Unless the period under herbicide is greatly prolonged it seems unlikely that major changes in the weed flora will be caused (Chapter 2).

## REFERENCES

ALDER, E. F., KLINGMAN, G. C. and WRIGHT, W. L. (1976). Herbicides in the energy equation, *Weed Sci.*, **24**, 99–106

ALLEN, H. P. (1975). ICI Plant Protection Division experience with direct drilling systems, 1961–1974, *Outl. Agric.*, **8**, 213–215

ATKINSON, D. and HERBERT, R. F. (1979). Effects on the soil with particular reference to orchard crops, from a Review of Long-term Effects of Herbicides, *Ann. appl. Biol.*, **91**, 125–129

BARKER, M. R. and WÜNSCHE, W. A. (1977). Plantio direto in Rio Grande do Sul, Brasil, *Outl. Agric.*, **9**, 114–120

BLEASDALE, J. K. A. (1963). Crop spacing and management under weed-free conditions, in *Crop Production in a Weed-free Environment*, 2nd Symposium of the British Weed Control Council, Blackwell, Oxford, p. 114

BULLEN, E. R. (1977). How much cultivation?, *Phil. Trans. R. Soc. Lond. B*, **281**, 153–161

CANNELL, R. Q. (1977). Soil aeration and compaction in relation to root growth and soil management, *Appl. Biol.*, **2**, 1–86

CANNELL, R. Q., DAVIES, D. B., MACKNEY, D. and PIDGEON, J. D. (1978). The suitability of soils for sequential direct drilling of combine harvested crops in Britain: a provisional classification, *Outl. Agric.*, **9**, 306–316

CHANCELLOR, R. J. (1979). The long-term effects of herbicides on weed populations, from a Review of Long-term Effects of Herbicides, *Ann. appl. Biol.*, **91**, 141–144

CUSSANS, G. W. (1976). The influence of changing husbandry on weeds and weed control in arable crops, in *Proceedings 1976 British Crop Protection Conference – Weeds*, pp. 1001–1008

DOWLER, C. C. and HAUSER, E. W. (1975). Weed control systems in cotton on Tifton loamy sand soil, *Weed Sci.*, **23**, 40–42

DREW, M. C. and SAKER, L. R. (1980). Direct drilling and ploughing: their effects on the distribution of extractable phosphorus and potassium, and of roots, in the upper horizons of two clay soils under winter wheat and spring barley, *J. agric. Sci., Camb*, **94**, 411–423

EDWARDS, C. A. and STAFFORD, C. J. (1979). Interactions between herbicides and the soil fauna, from a Review of Long-term Effects of Herbicides, *Ann. appl. Biol.*, **91**, 132–137

FRYER, J. D. and MAKEPEACE, R. J. (eds.) (1978). *Weed Control Handbook*, vol. 2, *Recommendations*, Blackwell, Oxford, p. 352

FRYER, J. D., LUDWIG, J. W., SMITH, P. D. and HANCE, R. J. (1980). Tests of soil fertility following repeated applications of MCPA, tri-allate, simazine and linuron, *Weed Res.*, **20**, 111–116

GREAVES, M. P. (1979). Long-term effects of herbicides on microorganisms from a Review of Long-term Effects of Herbicides, *Ann. appl. Biol.*, **91**, 129–132

GREEN, M. B. and McCULLOCH, A. (1976). Energy considerations in the use of herbicides, *J. Sci. Fd. Agric.*, **27**, 95–100

HAYWOOD, D. M., WILES, T. L. and WATSON, G. A. (1980). Progress in the development of no-tillage systems for maize and soya beans in the tropics, *Outl. Agric.*, **10**, 255–261

KEEN, B. A. and RUSSELL, E. W. (1937). Are cultivation standards wastefully high?, *J. R. agric. Soc.* **98**, 53–60

LAL, R., MAURYA, P. R. and OSEI-YEBOAH, S. (1978). Effects of no-tillage and ploughing on efficiency of water use in maize and cowpea, *Expl. Agric.*, **14**, 113–120

MAFF (1978). *Cereals without Ploughing*, ADAS Profitable Farm Enterprises, booklet 6, Ministry of Agriculture, Fisheries and Food, London, p. 84

OGBORN, J. E. A. (1972). The control of *Striga hermonthica* in peasant farming, *Proceedings of the 11th British Weed Control Conference*, pp. 1068–1077

ROBINSON, D. W. (1963). Crop situations where cultivations for weed control may be eliminated by use of herbicides: horticulture, in *Crop Production in a Weed-free Environment*, 2nd Symposium of the British Weed Control Council, Blackwell, Oxford, p. 114

THURSTON, J. M. (1969). Weed studies on Broadbalk, *Report Rothamsted Experimental Station*, 1968, pp. 186–208

TRIPLETT, G. B. and VAN DOREN, D. M. (1977). Agriculture without tillage, *Scientific American*, **236**, 28–33

VAN DER ZWEEP, W. (1971). Consequences of modern weed control for crop growing techniques, *PANS*, **17**, 20–25

WATERHOUSE, D. P. (1980). Growth regulators in sward management, *Proceedings of Conference on Weed Control in Amenity Plantings*, University of Bath, 1980

FURTHER READING

SCOTT RUSSELL, R. (1977). *Plant Root Systems: their Function and Interaction With the Soil*, McGraw-Hill, London, p. 298

# 9 ROW CROPS GROWN AS ANNUALS OR BIENNIALS

All the major food crops of the world as well as cotton and tobacco are grown as row crops. Most are planted as seeds although some are planted from stem tubers or other vegetative propagules. Young plants emerging from the soil soon face competition from weeds unless countermeasures are taken. Vegetatively propagated row crops, such as yams and potatoes, have an initial size advantage over most weeds, but their relatively slow emergence allows time for weeds to achieve a dominance over the crop. Thus effective control of weeds in row crops is essential if their yield potential is to be realised, and it is this group of plants that, worldwide, makes the greatest demands on human labour. The successful use of herbicides in row crops over recent years has been largely responsible for the changes in crop management (Chapter 8) and in the whole economic and social structure of farming that have occurred in Europe and other 'developed' areas.

## 9.1 PRINCIPLES OF WEED CONTROL IN ROW CROPS

In row crops, all but the most persistent perennial weeds are suppressed by thorough pre-planting or sowing cultivations, and it is annual weeds that become the major problem. The ability of the crop itself to suppress further weed growth depends on agronomic factors such as type and quality of crop seed, uniformity and density of planting and on the ability of the seedbed to supply adequate water and minerals. The incidence of pests and disease, especially at the seedling stage, can lead to a patchy stand of unequally sized plants in which continued control of weeds becomes difficult. However, the major factor in weed suppression by the crop is the nature of that crop. Those that are slow to achieve leaf cover, such as onions, are inherently poor competitors (Chapter 1). Potatoes, on the other hand, are able to make a dense leaf canopy which inhibits further weed growth. During the ripening stage of some crops, reduction of crop canopy as the leaves senesce once again allows weeds to thrive, and they may cause difficulties at harvest although direct losses from competition are probably slight.

The most competitive weeds are those that emerge at about the same time as the crop and both cultural and chemical methods have been devised for their control. One effective method of ensuring that the crop emerges into a clean, weed-free environment is known as the stale seedbed technique. This is achieved by delaying the sowing of the crop for some two or three weeks after seedbed preparation so that the first and most numerous flush of weed seedlings can be destroyed by spraying a fast acting herbicide such as paraquat immediately before the crop is drilled. A similar result is achieved by pre-emergence cultivation or herbicide application in crops, such as the potato, that are slow to emerge. In this situation herbicides are preferable to cultivation because they do not stimulate further weed germination.

### 9.1.1 Herbicides for row crops

Herbicides can be applied before planting or sowing and either before or after crop emergence, these treatments by convention being known as pre-planting, pre-emergence and post-emergence, respectively. Herbicides capable of killing perennial weeds or chemical soil sterilants such as dazomet, either of which would be lethal to the crop, can be applied pre-planting provided that sufficient time elapses for the residues to decompose or dissipate. Pre-emergence treatments may be of a foliar-acting contact type, e.g. paraquat, may act by root uptake from the soil, e.g. simazine, or may be taken up by weeds from leaves and soil, e.g. linuron. Contact herbicides, such as paraquat, destroy weed seedlings rapidly and leave no residue to damage the emerging crop, unlike soil-acting ones, but they lack the advantage of continued control of germinating weeds; mixtures of a contact and a soil-acting herbicide are therefore commonly used.

The first successful organic herbicides, 2,4-D and MCPA, were remarkable in that they exhibited such extreme selectivity between grasses and broad-leaved weeds when applied post-emergence. Although there are instances in which post-emergence selectivity has been found for row crops other than cereals (*Table 9.1*), most reliance must still be placed on pre-emergence treatment. Soil-acting and contact herbicides have nevertheless had a dramatic effect on production methods in several major crops including maize, cotton, soya

*Table 9.1*

*Post-emergence herbicides for some row-crop vegetables.*

| *Crop* | *Herbicides* |
|---|---|
| Cabbage | Alloxydim-sodium aziprotryne, desmetryne, propachlor |
| Carrot | Alloxydim-sodium linuron, metoxuron, prometryne |
| Onion | Alloxydim-sodium aziprotryne, ixoynil, methazole, pyrazone |
| Red beet | Alloxydim-sodium phenmedipham |

beans, brassica crops, legumes, sugar beet and potatoes. Atrazine, for example, provides season long control of many kinds of weeds in maize and sweet corn, and probably more atrazine than any other herbicide is applied today. No other herbicide combines the width of weed spectrum, prolonged persistence and crop safety as does atrazine in maize. Most other soil-acting herbicides that are sufficiently selective for use in row crops are less persistent and have a narrower weed spectrum. For prolonged weed control both mixed and sequential applications may therefore be required. Autumn sown onions in Britain, for example, may be sprayed during autumn, winter and spring with appropriate mixtures chosen from up to ten individual compounds, and even in the maize crop in which atrazine is generally safe and effective, many other herbicides can be used when atrazine resistant weeds are present or it is essential that a short persistence herbicide be used. The choice of herbicide may also depend on soil type, on the time of year, and sometimes on the crop cultivar being grown. Organic soils, for example, adsorb most herbicides so strongly (Chapter 6) that uneconomically high rates may be necessary for weed control.

Because most of the major crops and a majority also of weeds are contained within a relatively small number of plant families (Chapter 2) it is inevitable that closely related crops and weeds will occur together; when this happens difficult problems of chemical control are often encountered. Annual and perennial grasses have become the most troublesome weeds in cereals following general use of grass selective herbicides such as 2,4-D and similar situations already exist with weeds from Cruciferae in brassica crops and from Compositae in lettuce and chicory. The most difficult problem of all occurs when both crop and weed are of the same species as can happen with volunteer crops, annual beet and red rice. A large potential market for a herbicide with the extreme selectivity required for such problems will encourage efforts to find one, and already several herbicides capable of differentiating between Wild-oat and wheat or barley are in general use. However, compared with cereals, many row crops are minor ones in terms of total area of value, and these crops must therefore rely for their herbicides on developments initiated to satisfy larger and more lucrative markets. For example, the selective herbicides that can be used in red beet were all developed with the considerable worldwide production of sugar beet mainly in mind.

## 9.2 RECENT INNOVATIONS IN AGRONOMY AND WEED CONTROL

### 9.2.1 Pre-sowing seed treatments

The period between seed sowing and seedling establishment is a vulnerable one for sown crops. Cold, lack of water, attack by pathogens and pests, and high salt concentration from added fertiliser may all reduce seedling growth and survival; some crops, onions and carrots among them, germinate over an extended period, especially under adverse soil conditions. However, modern production methods (Chapter 8) may demand that every seed produces a plant, since failures leave gaps, which not only reduce total yield and allow weeds to grow, but also may ultimately affect the size grading of produce.

All seeds, and especially small ones, are sown within a few centimetres of the soil surface in a zone liable to rapid and extreme fluctuations in temperature and water status. The predicament of the crop seed and seedling, together with positive action that can be taken to give crop seedlings an 'edge' over their weed competitors has been elegantly described by Heydecker (1977).

Several methods of preparing seeds before drilling to enhance speed and uniformity of germination have been devised. Gardeners have long practised a pre-sowing soaking in water of peas and beans in the belief that this gives the seeds a good start; in fact the contrary is the case and after the initial hour or two longer soaking progressively retards subsequent growth. Two rather different approaches to shortening the time lag between sowing and establishment have been taken, namely 'fluid drilling' and 'priming'. In the former, developed at the National Vegetable Research Station in England, seeds are pre-germinated or chitted in water to the point where the radicle is beginning to emerge; to avoid damage to delicate and easily damaged seeds they are then drilled through a modified drill in a continuous ribbon of a gel similar to wallpaper paste (Currah, 1978; Salter and Darby, 1976). Faster and more uniform establishment of celery, parsnip and some other crops has been achieved, although differences visible at an early stage may disappear later. In the 'priming' method (Heydecker, 1974 and 1977) seeds are allowed to take up water from an osmotic solution such that the radicles reach the stage of being ready to emerge but are prevented from doing so by the osmotic concentration (water potential of about –12 bars) of the solution, what Heydecker called 'suspended germination'; several biologically inert compounds can be used as 'osmotic brakes' but most work has been done with polyethylene glycol '6000'. When the seeds are ready, after seven or more days in the osmoticum, they are dried enough to flow and sown immediately, whereupon germination is both rapid and uniform. In both fluid drilling and seed priming the growth of fungi and bacteria must be prevented and the seeds must receive enough oxygen.

Should either of these techniques come into general use, weed control methods will need adjustment, because the reduced time between sowing and emergence may reduce the crop selectivity of some pre-emergence herbicide/crop combinations. However, the improvement in crop growth and uniformity will reduce the period before the crop itself offers competition to weeds and weed control should be facilitated.

### 9.2.2 Precision drilling, monogerm seed and seed pelleting

The present tendency is for row crops to be sown at their final spacing or stations, so eliminating thinning and singling, both of which are labour intensive and technically difficult to do mechanically. Sowing to a station requires that seed be sown accurately (Chapter 8) which has led to the development of several types of precision drill which use belts with seed sized indentations at the desired spacing, each able to accept only one of a particular kind of crop seed. All types of beet produce an aggregate fruit containing several seeds; commercial seed was reduced in size mechanically ('rubbed seed') but latterly genetic monogerm sugar beet seed has been successfully introduced. Small awkwardly shaped seeds which would be difficult for precision drills to cope with can be rolled in a slurry of clay or other inert substance and formed into uniform small round pellets; not only does this facilitate drilling but pesticides or trace elements can be incorporated with the pellets.

These developments do little to aid weed control directly, and sowing to a station makes yet greater demands on herbicides to ensure a weed-free seedbed; however, drilling techniques that reduce the weight of crop seed used make it possible to sow expensive hybrid seed capable of producing more vigorous and competitive plants.

### 9.2.3 Activated carbon, protectants and mulches

Activated carbon adsorbs strongly many of the soil-applied herbicides used in row crops, thereby preventing them being taken up by plants (Chapter 6). Advantage can be taken of this property in transplanted strawberries by giving a pre-planting dip in a carbon slurry, but the method has not been very successful with row crops. Another way of using activated carbon is to lay a protective layer of carbon between the crop seed and the herbicide film at the soil surface, to prevent herbicide reaching the relatively deeply sown seed yet not prevent its being absorbed by shallow, rooted weeds. However, accurate placement of carbon is hard to achieve and in any case there is little advantage over using selective pre-emergence herbicides alone.

Another way of increasing herbicide selectivity is the use of specific chemical protectants or antidotes (Chapter 6), several of which are in use in maize and sorghum crops where they make it possible to use a herbicide giving some control of tropical nutgrasses (*Cyperus* sp.) in place of the standard atrazine treatment (Chapter 6). A combined protectant EPTC mixture is approved in Britain for sweet corn.

Polythene and petroleum mulches have found uses with certain crops. The former control annual and some perennial weeds and are being used on an increasing scale for vegetables in North America, for strawberries and for some tropical crops (Chapter 5). Petroleum mulch is primarily used for water conservation in arid areas and has little direct bearing on weed control although herbicides can be incorporated in the mulch.

## 9.3 INDIVIDUAL CROP CASE STUDIES

### 9.3.1 Potatoes

Until the late 1950s, satisfactory control of weeds in potatoes could be achieved largely by frequent cultivation before and after emergence, culminating in a final earthing up of the ridges when leaf growth prevented further operations. Chemical weed control was therefore regarded as an unnecessary extravagance. However, experience first in North America and later in Europe showed that excessive cultivations damaged roots and dried out the soil, leading to crop reduction, and in Scotland it was shown that cultivations for weed control when the soil was too wet could form the durable clods that cause difficulties for potato harvesters.

Initially the only suitable herbicide was dinoseb amine, which has a strong contact action on emerged weeds combined with a short residual soil activity. This was applied just prior to potato emergence, the potatoes having been planted and earthed up in a single operation, with no subsequent cultivation. Clod formation was reduced, but dinoseb alone was unable to provide prolonged weed control and

*Table 9.2*

*Some of the herbicides, including mixtures, in use in Britain in the potato crop*

| *Time of application* | *Herbicide* | *Target weeds* |
|---|---|---|
| Pre-planting | EPTC | Couch grasses |
| | TCA | Couch grasses and Wild-oat |
| Pre-emergence (contact) | Paraquat | Annual weeds, also checks perennial weeds |
| Pre-emergence (contact and residual) | Chlorbromuron | Mainly annual weeds |
| | Dinoseb in oil | |
| | Dinoseb-acetate + monolinuron | |
| | Linuron (alone and in mixtures) | |
| | Metribuzin | |
| | Monolinuron (alone and in mixtures) | |
| | Prometryne | |
| Post-emergence | Alloxydim-sodium | Couch grasses |
| | Barban | Wild-oat |
| | Metribuzin | Annual weeds |

weeds that became established soon after potato emergence competed with the crop. With the introduction of other contact herbicides, such as dinoseb in oil and paraquat, and of the soil acting ureas (linuron and monolinuron) and triazines (prometryne, trietazine) came reliable long-term weed control in the potato crop, without the need for any post-planting cultivations. Several other herbicides for pre-emergence control are now available; in addition EPTC and TCA are used pre-planting for control of couch grasses, and metribuzin can be used post-emergence provided the potatoes are no more than 15 cm high (*Table* 9.2) (MAFF, 1978).

### 9.3.2 Tropical root vegetables

The other major carbohydrate root crops, including yams (*Dioscorea alata*), sweet potatoes (*Ipomoea batatas*) and cassava (*Manihot esculenta*) have a production cycle similar to that of the potato (*Solanum tuberosum*), with a period of maximum susceptibility to weeds between emergence and the formation of a complete crop canopy. Chemical weed control in the potato has proved relatively easy, and no doubt the same would be true of the other root crops (Kasasian, 1971; Romanowski, 1977), but because most of the world production is by peasant farmers weeds are still controlled by hand (Chapter 1). Inter-planting with other crops, a common practice in the tropics, also makes herbicides difficult to use safely.

### 9.3.3 Brassica crops

Although there are several herbicides that can be used in drilled and transplanted brassica crops, there is still a need for herbicides with greater selectivity, especially when cruciferous weeds are prevalent. There are a few widespread weeds of cultivation closely related to *Brassica* and several others also within the family Cruciferae which have proved difficult to control selectively in brussels sprouts, cabbage, cauliflower, calabrese, swedes and turnips. For example, Charlock (*Sinapis arvensis*) and Field Penny-cress (*Thlaspi arvense*) are at best only moderately susceptible to the herbicides recommended for brassicas (*Table* 9.3) (MAFF, 1979).

Trifluralin incorporated thoroughly immediately following application will prevent growth of the more competitive weeds including Chickweed (*Stellaria media*) and Fat-hen (*Chenopodium album*) which may be less well controlled by other pre-emergence herbicides. Post-emergence herbicides cannot be safely applied until the crop has 3 or 4 true leaves, but annual weeds may well have passed their most susceptible stage by this time. In practice, therefore, combinations of hand and mechanical cultivation with both herbicide mixtures and sequential applications may be necessary, particularly when the crop occupies the land for a long period.

*Table 9.3*

*Herbicides that have been suggested for brassica crops. (In Britain carbetamide is applied to brassicas in winter months only.)*

| *Crop stage and method of application* | *Herbicide* | *Target weeds* |
|---|---|---|
| Pre-drilling | | |
| soil incorporation | Trifluralin | Annual species |
| foliar spray | Glyphosate | Perennial species |
| Immediately post-drilling | | |
| soil acting | Nitrofen | Annual species |
| Pre-emergence | | |
| soil acting | Propachlor | Annual species |
| Post-emergence when crop has several true leaves | Alloxydim-sodium | Grass weeds |
| mainly soil acting | Aziprotryne<br>Carbetamide<br>Propachlor | Annual species |
| mainly foliar-acting | Desmetryne<br>Sodium monochloracetate | Annual species |
| For use after transplating only | Chlorthal-dimethyl + methazole | Annual species |

*Table 9.4*

*Some herbicides suggested for carrots in Britain*

| *Crop stage and method of application* | *Herbicide* | *Target weeds* |
|---|---|---|
| 10 weeks or more pre-drilling | | |
| soil acting | TCA | Annual and perennial grasses |
| soil incorporation | Tri-allate | Wild-oat and annual grasses |
| soil acting | Trifluralin | Annual species |
| Pre-emergence | | |
| foliar and soil acting | Chlorbromuron<br>Linuron | Annual species |
| Post-emergence | | |
| foliar acting | Alloxydim-sodium<br>Dalapon | Annual and perennial grasses |
| foliar and soil acting | Chlorbromuron | Annual species |
| | Linuron | Annual species |
| | Linuron + metoxuron | Volunteer potatoes and annual species |
| | Prometryne | Annual species |

### 9.3.4 Carrots and other umbelliferous crops

In contrast to brassicas there are several wide spectrum herbicides to which Umbelliferae show considerable tolerance, both before and after emergence (*Table 9.4*); fortunately, also, there are no major arable weeds within this family. Before the present range of organic herbicides were discovered, certain oil fractions, including some types of TVO (tractor vaporising oil) were sufficiently selective to apply post-emergence for control of annual weeds in carrots. However, the use of oil always involved some risk of crop damage and taint when unsuitable grades were applied.

The relative ease of chemical weed control in carrots has led to the mechanisation of all stages of maincrop carrot production, which is now concentrated in a small area of eastern England with suitably deep, light soils. Beasdale (1963) (Chapter 8) showed how advantage could be taken of effective weed control to grow carrots at a closer and more productive spacing; not only have yields increased considerably, but by adjusting the spacing the close size specifications required by the processors can be met (Salter *et al.*, 1979).

Some of the herbicides now used so successfully in carrots, notably linuron, can also be used selectively for weed control in celery, parsley and parsnip.

### 9.3.5 Lettuce

Herbicides alone cannot yet provide a satisfactory control of all weeds. Pre-sowing or pre-planting-incorporated sprays of trifluralin can be followed by either chlorpropham, propham plus diuron or propyzamide, all pre-emergence; propyzamide may also be applied post-emergence. Sulfallate shows some selectivity pre-emergence but the margin of safety is said to be small. Weeds in the same Compositae family as lettuce, such as the Mayweeds (*Matricaria recutita* and *Tripleurospermum maritimum* subsp. *inodorum*) and Groundsel (*Senecio vulgaris*) are resistant and often build up in fields where lettuces are repeatedly grown. Early sown lettuce is a high value crop for which chemical soil sterilisation would be attractive, especially as some composite weeds carry the beet western yellows virus to which lettuce is susceptible (Chapter 1).

### 9.3.6 Onions and leeks

The problems faced by onion seedlings emerging into a weedy seedbed have already been described (Chapter 1); weed competition not only reduces total yields but the bulbs produced are likely to be small and thick-necked, yet weeds are hard to control by hoeing without damage to the small inconspicuous seedlings. Weed control is

*Table 9.5*

*Some herbicides suggested for drilled onions in Britain*

| *Crop stage* | *Herbicide* | *Target weeds* |
|---|---|---|
| Pre-emergence | | |
| foliar and soil acting | Pyrazone + chlorbufam | Mainly annual spp. |
| soil acting | Chlorpropham | |
| | Propachlor | |
| | Propachlor + chlorpropham | |
| Post-emergence | | |
| foliar acting | Dinoseb-acetate | Mainly annual spp. |
| | Ioxynil | |
| foliar and soil acting | Aziprotryne | |
| | Cyanazine | |
| | Ixoynil + linuron | |
| soil acting | Propachlor | |

also required throughout the life of the crop because it never produces a smothering canopy of its own.

Because seedling emergence is slow and erratic there is a need for a pre-emergence herbicide to prevent weed competition until the 'crook' stage has been passed, after which several selective post-emergence herbicides may safely be applied (*Table* 9.5). Good weed control is even more necessary on the autumn sown crop of Japanese onions, which remain in the soil from late summer until the following early summer. Indeed, autumn sown onion production would be uneconomic in Britain without effective herbicides.

Leeks, formerly grown only as a labour intensive transplanted crop, can now be sown to a stand and kept weed free by using much the same herbicide programme as for onions. Leeks are therefore no longer restricted to market gardens with access to labour but instead can be grown extensively with every phase of production mechanised.

### 9.3.7 Cotton

Cotton is a major cash crop wherever temperatures and water supply permit, including parts of North and South America, Africa, Europe, Asia and Australia; it is grown both on large scale plantations using the latest cultural techniques and on small peasant holdings. The crop is particularly prone to pest and pathogen attack and appropriate pesticide treatment is essential if high yields of good quality lint are to be obtained. Cotton is a poor competitor with other plants, especially after the sixth week from sowing, and yields can be severely reduced by perennial grasses, nutgrasses (*Cyperus* spp.) and other species. Weeds present at harvest may stain the lint, thereby reducing its value, and may also create harvesting problems. In large scale production herbicide use is now an accepted part of management, although in peasant farming where cotton may be one component of a mixed cropping system, herbicides are seldom used.

Trifluralin incorporated in the soil before planting, followed by the urea herbicide fluometuron pre-emergence, is often employed but alachlor (pre-emergence) provides better control of nutgrasses. The plant may then be left uncultivated until the closing over of the crop foliage restricts further weed growth, but in North America the scrambling vines, such as the Trumpet Creeper (*Campsis radicans*), may proliferate unless either cultivations or directed inter-row sprays of paraquat or glyphosate are applied from shielded sprayers. To date, there is no herbicide sufficiently selective to permit overall post-emergence spraying.

Repeated application of trifluralin over several years has decreased the prevalence of grasses but has increased nutgrasses and weeds in the cotton family (Malvaceae) such as Prickly Sida (*Sida spinosa*); other herbicides with a wider weed spectrum, and techniques for accurate placement of inherently non-selective weedkillers are therefore being tested (Hawtree, 1980).

## REFERENCES

BLEASDALE, J. K. A. (1963). Crop spacing and management under weed-free conditions, in *Crop Production in a Weed-free Environment*, 2nd Symposium of the British Weed Protection Council, Blackwell, Oxford, p. 114

CURRAH, I. E. (1978). Fluid drilling, *World Crops*, **30** (1), 22–24

HAWTREE, J. N. (1980). Weeds and cotton, *Outl. Agric.*, **10**, 184–190

HEYDECKER, W. (1974). Germination of an idea: the priming of seeds, University of Nottingham School of Agriculture Report for 1973–74, pp. 50–67

HEYDECKER, W. (1977). Seeds of success, *Sci. Hort.*, **28**, 100–115

KASASIAN, L. (1971). *Weed Control in the Tropics*, Leonard Hill, London, p. 307

MAFF (1979). *Weed Control in Potatoes, 1979*, leaflet RPG 22, Ministry of Agriculture, Fisheries and Food, London, p. 20

MAFF, (1979). *Chemical Weed Control in Vegetables, 1979*, leaflet HVG 28 (+ 1980 Supplement), Ministry of Agriculture, Fisheries and Food, London, p. 39

ROMANOWSKI, R. R. (1977). Integrated weed control in upland crops, in *Integrated Control of Weeds* (eds. J. D. Fryer and S. Matsunaka), University Of Tokyo Press, Tokyo, p. 262

SALTER, P. J. and DARBY, R. J. (1976). Synchronization of germination of celery seeds, *Ann. appl. Biol.*, **84**, 415–424

SALTER, P. J., CURRAH, I. E. and FELLOWS, J. R. (1979). The effects of plant density, spatial arrangement and time of harvest on yield and root size in carrots, *J. agric. Sci., Camb.*, **93**, 431–440

## FURTHER READING

Annual reports of research stations such as National Vegetable Research Station, Wellesbourne, Scottish Crops Research Institute, Invergowrie, Dundee and Weed Research Organization, Yarnton, Oxford. Current Ministry of Agriculture, Fisheries and Food bulletins and leaflets

EDDOWES, M. (1976). *Crop Production in Europe*, Oxford University Press, London, p. 318

FRYER, J. D. and MAKEPEACE, R. J. (eds.) (1977). *Weed Control Handbook*, vol. 1, *Principles*, Blackwell, Oxford, p. 510

FRYER, J. D. and MAKEPEACE, R. J. (eds.) (1978). *Weed Control Handbook*, vol. II, *Recommendations*, Blackwell, Oxford, p. 532

HEYDECKER, W. and COOLBEAR, P. (1977). Seed treatments for improved performance – survey and attempted prognosis, *Seed Sci. Technol.*, **5**, 353–425

MAKEPEACE, R. J. and HOLROYD, J. (1978). Weed control, in *The Potato Crop* (ed. P. M. Harris), Chapman and Hall, London, p. 730

MERCADO, B. L. (1979). *Introduction to Weed Science*, Southeast Asian Regional Center for Graduate Study and Research in Agriculture, Laguna, Philippines, 292 pp.

# 10 PERENNIAL CROPS OF ORCHARDS AND PLANTATIONS

## 10.1 WEED CONTROL STRATEGIES IN PERENNIAL CROPS

The herbaceous row crops, especially the cereals, pulses, root crops and vegetables, provide man with most of his food, but the many different kinds of perennials yield fruits and a great variety of other food and stimulant products, including spices, tea, coffee, cocoa, cola and hops, and manufacturing raw materials such as rubber and jute. In the underdeveloped world perennial crops are often grown in large plantations in which irrigation, fertilisers and pesticides are used efficiently, and are not subject to the constraints of peasant agriculture. In Europe and North America, also, the tendency is for perennial crops to be grown by specialists employing the most modern methods.

The planting and establishment of perennials, especially tree and bush crops, generally represents a large investment in labour and materials, made in the expectation of a long and productive period to follow. The life of the crop may be as short as two or three years for strawberries but it is often 50 or more years for some trees and bushes. The effects of weed competition in perennial crops are most severe when the plants are small, yet it is during this early and unproductive stage that crops are liable to be neglected because limited labour and pesticides are being deployed among profitable bearing plants. That this is false economy, and that young plants are particularly vulnerable to competition, has been repeatedly demonstrated in plantation crops. Intercropping in young perennial crops, although yielding some return, may cause similar retardation. As crops mature they produce more shade and may then effectively subdue further weed competition, this being particularly true of some tropical crops including rubber, cocoa and tea.

### 10.1.1 Weeds of perennial crops

The worst weeds in plantation crops are the perennial grasses and sedges which compete strongly with young trees. These weeds are most easily controlled before a perennial crop is planted; once the trees or bushes are in place it is no longer possible either to cultivate effectively or to apply overall the kind of herbicides needed for weed control, without damaging the crop. Herbicides that could be considered for pre-planting treatment of perennial weeds include glyphosate, amino-triazole, dalapon, 2,4-D and 2,4,5-T, the choice depending on the spectrum of weeds present. However thorough the pre-planting weed control some perennial weeds will survive and others will grow from dormant seeds or be brought in as seeds or vegetative propagules. Removal at an early stage when the number of plants is small is easily accomplished by physical removal or spot spraying but if neglected and allowed to multiply and spread they become progressively harder to cope with and may lead to premature grubbing of the plantation.

When good pre-planting control of the more pernicious perennial weeds has been achieved the chemical suppression of seedling weeds in the crop is comparatively simple; herbicides have now been successfully integrated into the management systems of most plantation crops (Chapter 8). Before the adoption of herbicides, plantation crops were either cultivated frequently or the weeds were grazed or cut back periodically.

When old trees have had to be destroyed before planting the crop it is important to ensure that they cannot support *Armillaria mellea* fungus, by exhausting the tree's food supplies; this is done either by tree girdling or by poisoning with herbicides such as ammonium sulphamate a year or two before the land is required (Kasasian, 1971; Fryer and Makepeace, 1978). This fungal disease can strike anywhere, but is more common in tropical crops such as tea and cocoa.

Alternatively, a cover crop of a scrambling legume such as *Pueraria phaseoloides* can be grown in the tropics, or of grasses and clover in temperature regions. Cover crops provide a better surface than a frequently tilled one for men and machines, they protect the soil from erosion and in very wet districts transpire surplus water, but under drier conditions they can become too competitive with the crop, restricting root run and severely reducing water and nutrient uptake.

Experience with soils kept weed-free chemically (Chapter 8) shows that they combine some of the advantages of clean cultivation and cover cropping. However, growers have sometimes been reluctant to sacrifice the advantages of a cover crop, and many have adopted an alternation of herbicide-treated strips containing the crop with cover-cropped alleyways, despite the fact that trees have often been shown to grow faster and yield better in the complete absence of other plant competition.

Mulches of organic waste material or black polythene can be placed around perennial plants to conserve water, protect the soil from erosion and control weeds. Mulches of organic waste do not prevent all weed growth and should therefore be used in combination with herbicides (Davison and Bailey, 1979) (Chapters 5 and 11).

### 10.1.2 Herbicides for perennial crops

Perennial plants have been found to possess considerable intrinsic resistance to many herbicides (Chapter 6) and their sheer size apparently reduces susceptibility to herbicides that are lethal to seedling crops and weeds.

The first effective herbicides used selectively in tree crops were various phytotoxic oils and sodium arsenite, both of which destroy the aerial parts of herbaceous plants but provide little long-term control of perennial species. Sodium arsenite has the further disadvantages of being highly toxic to mammals and of causing an accumulation of arsenic in the soil after repeated applications.

The many herbicides now available are capable of ensuring an adequate degree of weed control in perennial crops but some perennial species continue to be a problem. For example, the tuber-forming *Cyperus* and bulb-forming *Oxalis* species can only be fully controlled by soil sterilisation before planting, but this is too expensive for most crops. Once the crop has been planted the weeds with rapidly extending root and rhizome systems such as Field Bindweed (*Convolvulus arvensis*), Creeping Thistle (*Cirsium arvense*) and rhizomatous grasses are difficult to control, especially in herbaceous crops such as strawberry and asparagus. Overall treatments of crop and weed are possible in some instances, and applications of MCPB in this way can be made after harvest to control Field Bindweed in blackcurrants and gooseberries; in addition, propyzamide applied overall during the dormant season and alloxydim-sodium during the growing season control some grass species while showing good selectivity towards many kinds of perennial crops. However, overall application of most other herbicides that exert a useful control over perennial weeds would damage the crop.

In the absence of perennial weeds, soil-acting herbicides including carbamates, ureas, triazines, uracils and oxadiazon can be used to control germinating weed seedlings; oxadiazon also controls bindweeds but application must be made well before the crop begins growth in the spring, and several annual applications may be needed for an adequate control. The choice of pre-emergence herbicide will depend on its degree of toxicity to the crop being grown, its cost and the predominant weed flora. Periodic revision of the herbicide or mixture will prevent the build up of weeds resistant to any single compound. Individual weeds that survive and multiply despite herbicide treatment can be spot sprayed with a foliar-acting compound such as paraquat (*Table 10.1*). Because individual crops have shown sensitivity to certain herbicides only recommended herbicides should be used; perennial crops often represent a large investment and experiments

*Table 10.1*

*Herbicides used for weed control in perennial crops. (Other herbicides are used for specific crops and not all those listed are universally suitable.)*

| *Stage of crop applied* | *Herbicide* | *Target weeds* |
|---|---|---|
| Pre-planting | | |
| chemical sterilant | Methyl bromide<br>Dazomet | Seed and perennating organs of perennial species |
| soil acting | TCA | Perennial grasses |
| foliar acting | Dalapon | Annual and perennial grasses |
| | Glyphosphate or aminotriazole | Wide range of perennial species |
| | 2,4-D or MCPA | Broad-leaved annuals or perennials |
| | 2,4,5-T | Woody weeds and broad-leaved herbaceous perennials |
| Post-planting | | |
| soil acting, directed away from crop whenever possible | Atrazine, chloroxuron<br>Chlorpropham<br>Chlorthal-dimethyl<br>Chlorthiamid, dichlobenil<br>Diuron, lenacil, oxadiazon<br>Propachlor, propham<br>Simazine | Germinating weed seedlings |
| spot application – generally must be away from crop | Alloxydim-sodium<br>Aminotriazole<br>Asulam<br>Bromacil, chlorthiamid<br>Dichlobenil, 2,4-D<br>MCPA<br>Glyphosate<br>Paraquat | Perennial weeds and herbicide-resistant annual species |

on commercial holdings with new herbicides involve a risk of long-term crop reduction.

The herbicides sometimes used in perennial crops are listed in *Table 10.1* in three categories, namely for perennial weed control before planting, for maintenance of a weed-free condition in established plantings and thirdly for spot treatment of surviving weeds.

## 10.2 HERBACEOUS TEMPERATE CROPS

### 10.2.1 Strawberries, rhubarb and asparagus

In these formerly labour intensive crops herbicides have enabled a change to more extensive and partly mechanised culture to be made, generally with a minimum of cultivation. Herbicides now form an integral part of the husbandry although perennial weeds can be difficult to control and may cause problems.

Strawberry fruiting plantations and runner beds are notoriously hard to keep clean especially when grown under polythene tunnels, and weeds can soon choke the plants, reducing yield and making picking difficult. Initially, chemical control was attempted with mixtures based on 2,4-DES and chlorpropham, but control was often inadequate and always short-lived. Simazine, with its wide weed spectrum and relatively long persistence, appeared at first to be ideal for the crop, but it was soon found to cause crop damage, especially from spring application, and recommendations for its use are now subject to some limitations. However, several other herbicides show selectivity to strawberries (Clay, 1980), and herbicide mixtures and sequential combinations now make season-long weed control possible (*Table 10.2*) (MAFF, 1979). Lenacil has proved to be much safer than simazine but

*Table 10.2*

*Herbicides used for weed control in strawberries in Britain. (The use of some of the herbicides in the table is subject to certain reservations – always consult current recommendations before application.)*

| *Stage of crop applied* | *Herbicide* | *Target weeds* |
|---|---|---|
| Pre-planting | Aminotriazole | Perennial grasses |
| | Asulam | *Rumex* spp. |
| | 2,4-D | *Convolvulus* spp. |
| | Glyphosate | Perennial weeds |
| | Paraquat | Annual weeds |
| | TCA | Perennial grasses |
| | Trifluralin – followed by incorporation | Annual weeds |
| Post-planting | | |
| Soil-acting overall treatment | Chloroxuron, chlorthal-dimethyl, 2,4-DES, diphenamid, lenacil, pendimethalin, propachlor or simazine, alone or in mixtures | Annual weeds |
| | Ethofumesate | *Trifolium repens* |
| Foliar-acting overall treatment | Alloxydim-sodium | *Agropyron repens* |
| | 3,6-Dichloropicolinic acid | *Cirsium arvense* |
| | Paraquat | Annual weeds and runners (see text) |
| | Phenmedipham | Small weed seedlings |
| Spot or local treatment | 2,4-D amine<br>Glyphosate<br>Terbacil | Patches of perennial weeds |

Some cultivars also show tolerance of oxadiazon as a spot treatment of *Convolvulus arvensis* in spring, or in autumn for simazine-resistant annuals.

because of its relative expense cheaper substitutes are often considered. There are few examples of selective foliar-acting herbicides in perennial crops, but several, including some first developed for beet, can be applied to strawberries (*Table 10.2*). Fruiting strawberries are often burnt or cut back to the crowns after harvest, and at this time a directed spray of paraquat can be used to destroy both small weeds and unwanted runners in the inter-row. Effective soil-acting herbicides such as simazine and lenacil create conditions favourable for perennials including Creeping Thistle and Field Bindweed, and these need prompt physical removal or herbicide treatment (*Table 10.2*) to prevent their spread.

Rhubarb tolerates several herbicides during the dormant season, including simazine, dalapon and TCA, and these can provide long-term freedom from annual weeds and grasses. Control of broad-leaved perennials such as docks and nettles can be accomplished by careful spot treatment but some crop damage may be unavoidable.

Asparagus can also be treated with herbicides when it has died down completely in the winter. Soil-acting materials that have proved safe in this crop include lenacil, monuron, diuron and simazine, and terbacil and dalapon have been used for spot-treating perennials. Advantage can also be taken of the dormant period before the spears emerge to kill any surviving weeds with paraquat or other non-persistent contact herbicide. A weed of asparagus that is difficult to control is the asparagus itself – seedlings from seeds spread by birds tolerate most of the soil-applied herbicides.

## 10.3 HERBACEOUS TROPICAL PERENNIALS – SUGAR CANE AND BANANAS

### 10.3.1 Sugar cane

Cane is grown over a range of tropical and subtropical climates. The period between successive crops, or ratoon crops, can vary from less than 12 months in Louisiana, USA, to almost two years in Hawaii. The initial propagation is by 2 or 3 bud sections of the upper part of the stem discarded during harvesting, which are planted about 4–6 cm deep in rows some 2 m apart. After the first crop, a planting may well produce upwards of 10 ratoon crops; yields tend to fall as disease spreads or the soil becomes compacted. Under irrigation in the tropics cane can produce a dense leaf canopy within four months of planting although the period will be longer in drier and cooler areas. The most vulnerable period for weed competition is from a few weeks after emergence until the leaf canopy is complete, and in the past the rapid growth of tropical annual weeds, perennial grasses and nutgrasses (*Cyperus* spp.) has required repeated hand weeding. Cane, however shows tolerance of many herbicides including atrazine, simazine, diuron, 2,4-D and MCPA, ioxynil, asulam and dalapon. Even paraquat is sometimes recommended for young cane which, although scorched, is apparently not permanently harmed.

Wherever cane is grown on a large scale as a cash crop for sugar export, chemical weed control using large ground machines or aircraft has become standard practice. But where labour is readily available a combination of hand work and herbicide application by knapsack sprayers is more common. However, peasant production using hand tillage only still accounts for much of the cane produced locally for small scale processing.

A recent development has been the use of herbicides, not for killing weeds but as chemical ripening agents or to prevent tasseling (flowering). Flowers can form in some cultivars grown for a long season, as in Hawaii, and flower prevention by spraying with diquat at very low rates has increased yields. Natural ripening of sugar cane can be promoted by several compounds, including the herbicides glyphosate and asulam, leading to an increase in the amount of extractable sucrose in the cane. Chemical ripening and anti-tasseling compounds have to be applied to the dense standing crop from the air (Nickell, 1976).

### 10.3.2 Bananas

Bananas are susceptible to weed competition when young but if sufficiently densely planted mature plants then exert control by shading. Cover crops have been used but the most popular one, of *Commelina* spp., can act as a host for nematodes and for viruses causing disease in banana (Kasasian, 1971). Cultivation, which damages the superficial root system, and slashing with machetes are the usual methods of control but they are expensive and not very effective, and herbicides now form an attractive alternative. Several, including atrazine, simazine, linuron and ametryne, have been suggested either alone or in combination with paraquat. Paraquat, however, must be kept away from crop foliage. Generous application of hormone weedkillers such as 2,4-D can cause sufficient distortion of the banana growing point to make the pseudostems snap off at ground level, but carefully directed sprays on non-volatile formulations of 2,4-D or MCPA have been used to control scrambling *Convolvulus* spp. in banana.

## 10.4 WOODY PLANTATION CROPS OF THE TROPICS AND SUBTROPICS

### 10.4.1 Strategy of weed control

The great richness of the tropical flora is reflected in the variety of woody species grown for food, timber and other purposes, including and pimento as well as oil palm and coconut and teak. Most of these crops such as tea, cocoa and coffee, various citrus, mango, avocado woody crops are grown in well-managed plantations, which have been established from tropical rain forests during the past 100 years.

Unprotected soil surfaces are liable to erosion and therefore cover crops are frequently grown. These may be temporary, but are often permanent in more widely spaced tree crops. Legumes of various kinds are used, such as *Pueraria phaseoloides*, which do not grow to any great height and have the added advantage that their roots bear nitrogen fixing nodules. However, cover crops must be chosen and managed to avoid undesirable crop competition. Cash crops, including soya bean, maize or bananas, may be grown in the alleyways of young plantations, but inter-cropping is labour intensive and is therefore more common among peasant farmers than in large commercial plantations.

A further consideration in irrigated crops is the method of irrigation. Methods that wet the whole soil surface provide ideal conditions for weed growth, but trickle systems may well allow weed germination in the immediate vicinity of the source only, the constant superficial dryness elsewhere effectively preventing weed seedling establishment.

Weeds, especially perennials, can rapidly establish a stranglehold on woody plantations in the tropics. Formerly, this was prevented by frequent cultivations, sometimes coupled with cover cropping, but cultivation favours erosion, leads to loss of organic matter and increases loss of soil minerals by leaching (Seth, 1977). Chemical control, which would control the worst weeds without the problems created in tropical climates by cultivation, was therefore welcomed. Before modern herbicides became available, phytotoxic oils were used in citrus and sodium arsenite in rubber, but these have largely been replaced by the safer, cheaper and more effective herbicides that now control both germinating seedlings and the more competitive perennial grasses and sedges. Woody plants tolerate soil-acting herbicides better than most other crops and several compounds have been proved safe after prolonged trials; one or more of the wide spectrum, long-persistence herbicides (Chapter 6), is now recommended for virtually every major woody crop. In addition, essentially non-selective foliar-acting herbicides such as paraquat and glyphosate can be used for spot-spraying or directed application.

A further weed problem of hot climates is created by epiphytic and parasitic flowering plants and by epiphytic mosses and lichens. Epiphytes on oil palm, for example, can be controlled by directed sprays of 2,4-D or ametryne; mosses and lichens on cocoa can be dislodged by spraying with copper oxychloride. Parasitic plants are more harmful and are often more difficult to control, although chemicals sufficiently selective have been found for some crops (Chapter 4) (Seth, 1977).

### 10.4.2 Coffee, cocoa and tea

Coffee and cocoa are susceptible to weed competition when young, and bearing plantations can also suffer considerable yield losses from weeds. Coffee benefits from being planted under shade from existing trees, which reduces the growth and competitiveness of the predominant perennial grasses. Cover crops, generally of legumes, are often grown under new plantings, although yield reductions due to some cover species have been demonstrated, particularly where water is scarce. For cultivations to control weeds properly they must be repeated frequently and other methods such as mulching and herbicides have been used where labour is expensive.

Simazine was the first recommended herbicide that provided prolonged weed control but many other herbicides including dalapon, diuron and paraquat have been used successfully against locally prevalent or resistant weed species. Both tea and coffee rely for their market value on the subtleties of aroma and flavour and any suggestion of taint due to pesticides must be avoided; fortunately only the now outmoded PCP has been linked with taint in coffee.

Cocoa is often grown in areas with no dry period, consequently weeds grow vigorously throughout the year. Young cocoa plants are sensitive to competition to the extent that some plants may die when weeding is delayed. Shade, either from existing large trees or from the crop itself as it matures, is therefore important in reducing weed competition. Traditionally, weed control is by repeated hand slashing with machetes but this merely provides temporary control of weeds capable of rapid re-growth and the machetes often damage the crop plants. Diuron and atrazine have been recommended for controlling germinating weeds, while foliar-applied paraquat or glyphosate are used for controlling perennial grasses.

Well-established tea provides a dense shade that prevents the growth of most weeds but young tea, and older plantations that have just been pruned, allow weeds an opportunity to grow. Before planting, the perennial grasses, the bane of almost all tropical crops, should be eliminated either culturally or chemically; young plants in the nursery are grown in sterilised soil to ensure that no competition is suffered at that stage. Cultivation in planted out tea inevitably damages roots, apparently increasing the risk of *Armillaria mellea* infection which is always a hazard in newly planted tea. A combination of paraquat and simazine has generally proved sufficient to keep the bushes weed free until they become large enough to smother further weed growth, but other herbicides are being tested for particular local problems (Rahman, 1975). Cover crops are sometimes grown and according to Kasasian (1971) these include the bulbous *Oxalis latifolia* and *O. corymbosa*, which in other situations are pernicious weeds, although Kasasian does give a warning that subsequent planting of another crop on a former tea garden cover cropped with *Oxalis* spp. might prove difficult!

### 10.4.3. Rubber and sisal

Rubber, which is indigenous to the rain forests of the Amazon basin is now grown commercially in similar situations in Indonesia, Malaysia and the former Indochina. Perennial grasses such as Lalang (*Imperata cylindrica*) compete strongly with young trees under these conditions and should therefore be eliminated before planting. Formerly, sodium arsenite was generally used for grass control but this has been largely superseded by dalapon, paraquat, glyphosate and the organic arsenicals MSMA and DSMA. The same herbicides can also be used as directed sprays in established plantations.

Before herbicides were introduced the area immediately adjacent to the trees was cleaned by hand, sometimes aided by organic mulching. Increasingly, however, hand labour is being replaced by a combined paraquat and residual herbicide application to the strip containing the trees. Because paraquat can damage any exposed leaves or green stems the spray should always be a directed one.

Cover crops of legumes are sometimes recommended for rubber although their eventual decay may provide the trees with excessive nitrogen (Kasasian, 1971). As the shade increases a cover crop of ferns is sometimes established. If there is no cover crop then the weeds in the alleyways are cut or sprayed occasionally, but the strip containing the trees must be kept weed-free for access. A weed problem that arises in plantations of trees old enough to shed seeds is the germination of many unwanted young rubber trees; these are resistant to most of the soil-acting herbicides but can be killed by spot spraying with 2,4-D or 2,4,5-T.

Rubber has recovered its former importance in world trade by remaining competitive. Greatly increased productivity has followed the use of improved cultivars and of ethylene to stimulate latex flow; good weed control using herbicides is an integral part of managing modern high yielding rubber plantations.

Sisal, although a crop of relatively dry climates compared with rubber, also suffers most from weed competition during its first two or three years from planting. Cover crops of Kudzu (*Pueraria* spp.) keep out weeds and may benefit the crop, but in general cover crops, like weeds, appear to reduce the growth of the young plants. However, sisal is resistant to several of urea, triazine and uracil herbicides, which simplifies chemical weed control; it is said to be susceptible to 2,4-D, dalapon and other translocated herbicides unless the sprays are carefully directed. The most reliable long-term weed control is achieved with bromacil, but another herbicide should be substituted during the last few years of a plantation's life to prevent damage to susceptible succeeding crops from bromacil residues.

## 10.5 FOOD AND FRUIT CROPS OF THE TROPICS AND SUBTROPICS

These include citrus, avocado, mango, coconuts and oil palm as well as a variety of other fruits, nuts and spices grown on a smaller scale.

Both cover crops and weeds, particularly grasses, are considered too competitive to be tolerated in citrus in most countries and some form of weed control such as mulching or clean cultivation is usually necessary. It was therefore in citrus that erosion and other problems associated with clean cultivation prompted application of high aromatic content herbicidal oils (Chapter 6). These were sprayed overall, but avoiding the trees, at about 400 l/ha; however, because there was rapid regrowth of perennial weeds, repeat applications every six weeks were required. Citrus, like many other tree crops, is tolerant of urea, triazine and uracil herbicides applied to the soil; terbacil and bromacil, which provide a wide spectrum of control of annual and perennial weeds over a prolonged period, are now often used in preference to the herbicidal oils. Dalapon, applied for grass control, can sometimes be leached to the crop rooting zone in sufficient quantities to cause leaf and fruit drop. In wet areas epiphytic mosses, ferns, bromeliads and orchids on the trees may require control using lead arsenate or cuprous oxide (Kasasian, 1971).

In common with many other woody crops, coconuts suffer most from both cover cropping and weed competition when the trees are young, but the roots of young trees may also be damaged by cultivation. Young trees are often neglected at this early and vulnerable stage just because they are still unproductive. The herbicides dalapon, diuron, simazine, atrazine and paraquat have all been used successfully in coconuts; herbicides are expensive, and their use may be uneconomic unless the trees are planted sufficiently closely together. By no means all herbicides are safe and both bromacil and 2,4,5-T have damaged crops; the symptoms of damage may take several years to appear, which underlines the need for full and careful testing under the local conditions of soil and climate before any firm recommendations are made. Large bearing trees no longer suffer so severely from competition and chemical control may no longer be economically justified but control of the sward under the trees by cutting or livestock grazing is still required to facilitate the harvesting of fallen nuts.

## 10.6 TEMPERATE WOODY FRUIT CROPS

Uncontrolled weeds not only reduce crop vigour and yield (Chapter 1) but make picking of fruit, pruning and pesticide application difficult; in the past weeds have generally been controlled by combinations of hand or mechanical cultivation, mulching or by cutting or grazing grassed cover crops. Because these methods are not wholly effective and involve expensive hand labour the chemical control of weeds was welcomed by growers. The effects of herbicides on both weeds and crop have been intensively investigated, especially in Europe and North America since the early 1960s and there are now detailed recommendations for their safe use (MAFF, 1977; Fryer and Makepeace, 1978). Overall chemical weed control has been generally adopted in raspberries, blackcurrants, gooseberries and grape vines; in tree fruits a partial grass or other cover in the alleyways is often combined with a herbicide-treated, weed-free strip containing the trees. As in all perennial crops the most difficult weeds to control are the perennial species such as Bindweed (*Convolvulus arvensis*), Common Couch (*Agropyron repens*) and Creeping Thistle (*Cirsium arvense*); in the absence of competition from annual weeds following the use of herbicides, more unusual perennial species including tree seedlings, Hogweed (*Heracleum sphondylium*) and asparagus may also thrive.

### 10.6.1 Bush and cane fruits

Since the introduction of herbicides the yields of blackcurrant have increased. This crop appears to grow better on a non-cultivation system, which allows its roots to ramify near to and even above the soil surface (Chapter 8); previously, the superficial roots were systematically destroyed by the frequent cultivation that was necessary to control weeds. Some species of moss are sufficiently herbicide resistant to survive and in all but the driest regions they provide a natural ground cover between the bushes. Herbicide programmes are based on paraquat and simazine, but the appearance of resistant weeds may require sequential or spot application of clorthiamid or dichlobenil (as granules) or of lenacil, propyzamide, glyphosate or MCPB. Some annuals, including Knotgrass (*Polygonum aviculare*) and Common Orache (*Atriplex patula*) are moderately resistant to simazine but their build up can be prevented by occasional use of diuron or lenacil in place of simazine. If paraquat is used in blackcurrants then it must only be applied as a carefully directed spray because the buds never become wholly resistant even during winter; in gooseberries the winter buds are more resistant. Paraquat may be used to destroy winter annuals and some perennials such as Creeping Buttercup (*Ranunculus repens*) before the annual application of soil-acting herbicides in early spring. The benzonitrile herbicides chlorthiamid and dichlobenil are suggested for late winter application to bushes established for a year or two; both will control germinating annual weeds but their principal use is as a spot treatment for perennial broad-leaved species. Common Couch (*Agropyron repens*) is resistant to simazine and no more than checked by paraquat, but a winter application of propyzamide will control couch grasses, annual grass seedlings and Common Chickweed (*Stellaria media*). Blackcurrant and gooseberry tolerate MCPB in September when all extension growth has ceased, and this herbicide can be used as an overall spray for controlling Bindweed or as a directed spray for control of Creeping Thistle and Creeping Buttercup. Despite skilful choice of herbicides, the more intractable perennial species may survive and multiply. The only practical solution may be hand removal of seedling trees, blackcurrants, or other herbicide tolerant crops such as asparagus, or of resistant herbaceous perennials including Hogweed. However, many of these plants may be killed with the total herbicide glyphosate applied as a carefully directed spray or through a herbicide glove, a technique devised for chemical roguing for Wild-oat (*Avena spp.*) in cereal crops; application may also be made by brush, using a gel formulation. Planned application of various herbicides during the lifetime of a fruit plantation enables the grower to achieve weed control at the lowest possible cost by using mainly the cheaper materials such as simazine and paraquat, only resorting to the more expensive materials when the appearance of resistant weeds demands them. Application rates of soil-acting herbicides may be reduced in the last year or two of a plantation's life, and the use of the more persistent compounds such as the benzonitriles should be avoided altogether at this time, to prevent the accumulation of residues that might damage a succeeding crop.

Cultivation during the growing season inevitably damages the young shoots of raspberries, damage that can now be avoided with chemical weed control. Annual application of simazine was rapidly adopted as a standard treatment, and coupled with this was a change from the frequent cultivations of the past to a totally non-cultivation system throughout the ten or more years' life of a plantation. Early experience in Scotland showed that initially insignificant numbers of perennial weeds take advantage of the sudden lack of competition from annual weeds, rapidly build up and create a weed problem far worse than that originally present. For example, raspberry plantations totally overrun by thistles (mainly *Cirsium arvense*), docks and sorrels (*Rumex* spp.), Common Couch (*Agropyron repens*) and Colt's-foot (*Tussilago farfara*) were not uncommon a year or two after the introduction of simazine. Atrazine, bromacil or lenacil may now be used instead of simazine to control annual weed seedlings. Atrazine has much the same soil activity as simazine but also has foliar activity against emerged weeds; bromacil also has leaf and soil activity against annuals, but in addition has some useful activity against perennials. The advantage of lenacil lies in its wider weed spectrum which includes several annual species not fully controlled by simazine. As with the bush fruits, weed control programmes are devised making the maximum use of relatively inexpensive herbicides (MAFF, 1977).

Reduction in both cultivation and weed competition may lead to excessive abundance of vigour of new shoots and some control of their growth may then be necessary (Lawson and Wiseman, 1979).

### 10.6.2 Grape vines

Grapes for wine production are one of Europe's most ancient crops and old manuscripts and stained glass windows frequently depict men cultivating the soil around vines (*Figure 10.1*). That grapes grow

**Figure 10.1** Fifteenth-century illustration of soil cultivation in a vineyard.

perfectly well without cultivation was demonstrated by experiments early this century in which the soil around the vines was partly covered by concrete slabs (Roques, 1976), but the continued need for weed control by frequent soil disturbance has until recently prevented non-cultivation methods in practice. In some areas weeds in winter are considered beneficial because they increase the loss of water from the soil but in summer tangled stems of Bindweed (*C. arvensis*) can strip the fruit from the plants and perennial grasses such as Common Couch (*A. repens*) in colder regions, and Bermuda-grass (*Cynodon dactylon*) in the warmer areas, compete strongly with the crop for water and nutrients. It is also widely believed that weeds create an environment favourable for the spread of downy mildew in the crop. To cope mechanically with annual and perennial weeds demands thorough and drastic cultivation with resultant damage and actual loss of plants, and in many vine areas, therefore, herbicides have replaced traditional methods. The vines suffer less physical damage above ground, the roots can actively explore the fertile surface layers formerly denied them and the non-cultivated soil may give up more heat during a cold spring night thus reducing the risk of frost damage to flowers. Using simazine or oxidiazon, and paraquat, most annual weeds can easily be controlled but perennial grasses and Bindweed (C. *arvensis*) remain problems.

One successful method of overcoming both annual and perennial weeds in the crop rows is to plant into a continuous strip of black polythene which is then left in place for three or four years; the alleyways meanwhile are kept clean by superficial cultivation. Not only is weed control simplified but the crop is said to come into bearing a year earlier (Roques, 1976), which more than compensates for the extra costs. On the steep, shallow soils of Beaujolais and Alsace in France, temporary cover crops of Italian ryegrass are sown annually in the summer to protect the soil from erosion in winter; in the following spring the ground cover is killed with paraquat to prevent competition with the crop for water during the growing season.

The quality of wine made from grapes depends on delicate subtleties of flavour, and growers are therefore reluctant to apply any pesticide that might impair its quality. Fortunately the herbicides at present used do not appear to produce any such undesirable effects in either the harvested grapes or in the wine produced, although cultural control of weeds by hand and machine is still preferred in the areas producing the highest quality wines.

### 10.6.3 Temperate tree crops – apples, pears and stone fruits

Current methods of fruit production stress the need for a rapid financial return from a heavy investment in trees, land and equipment; this requires a management system that minimises competition from weeds or cover crops, allows ready access for picking and other operations, prevents soil erosion and maintains soil structure and fertility. When larger more widely spaced trees on vigorous rootstocks were in vogue, the weeds or cover crop were manipulated to provide sufficient competition to control tree growth but this is now seldom necessary. The orchardist must choose between the possible combinations of clean cultivation, cover cropping or freedom from weeds maintained with herbicides (Robinson, 1976). Clean cultivation accelerates loss of soil organic matter and reduces root exploitation of the cultivated zone; furthermore, cultivated soils provide a soft and uneven surface for pickers and machinery. Because of these disadvantages many orchards were sown to grass mixtures of Timothy

(*Phleum pratense*), Perennial Rye-grass (*Lolium perenne*) and White Clover (*Trifolium repens*). Grassing of orchards was usually delayed for several years to reduce the severity of the growth check that occurs and, in order to prevent excessive crop competition, the sward is kept below about 10 cm by regular cutting. Chemical alternatives to regular mowing have been evaluated, including maleic hydrazide (Stott, 1972) and mefluidide (Bushong *et al.*, 1976). Chemical retardation is practicable but is not at present economically attractive in fruit production although it is being adopted for controlling urban roadsides (Waterhouse, 1980). Grass sward makes an excellent surface for all orchard operations and the fruit produced is often of superior keeping quality, yet it does compete with the crop because relatively few tree roots can exploit the fertile superficial layers under the grass (Atkinson, 1977). In a dry year the loss of water is appreciably greater from a grass sward than from either a cultivated surface or one kept weed free with herbicides. In addition, taller grasses and weeds adjacent to the trunks of young trees are very competitive and provide shelter for fungi and rodents that damage bark.

At Long Ashton Research Station in England experiments with cover crops of clover (*Trifolium* spp.) have been promising; clovers do not require frequent cutting, appear to lose less water than an equivalent grass sward and atmospheric nitrogen is fixed in their root nodules. The preferred species at Long Ashton is Strawberry Clover (*Trifolium fragiferum*) because its flowering period is late enough to avoid attracting bees to the orchard when insecticide residues will be present.

Apples and pears tolerate simazine and diuron at rates above those necessary to kill weed seedlings and other herbicides including chlorthiamid, dichlobenil, oxadiazon, terbacil and bromacil have been used for spot treatment of resistant weeds. Stone fruits may show foliar symptoms of herbicide damage and may be less tolerant but no yield reductions have been reported; however, herbicide applications should be strictly in accordance with local recommendations. Paraquat, aminotriazole and glyphosate will damage fruit tree foliage and green bark but directed sprays can be applied to weed foliage. Suckers of stone fruits have been killed with glyphosate without the herbicide being translocated to the parent tree.

Complete control of weeds without cultivation is now practised in some orchards. The trees grow more vigorously, come into bearing sooner and crop more heavily, producing larger fruits (*Table 10.3*). A higher incidence of storage rots caused by infection with *Phytophthora syringae* has been attributed to splashing of water drops containing fungus spores from bare soil to the fruit, but it has been found that there is in any case a greater incidence of storage rots due to *Phytophthora* in larger fruits, irrespective of the plantation system

*Table 10.3*

*Mean annual yield fruit (tonnes/ha) from different apple management systems at Long Ashton, 1973–1977. (From data supplied by K. Stott.)*

| *Treatment* | *Cox* | *Golden Delicious* |
|---|---|---|
| Trees in herbicide treated weed-free strip, alleyways grassed | *12.2* | *20.4* |
| Overall clover | *13.5* | *24.8* |
| Overall herbicide treated and weed free | *14.7* | *26.3* |

that produced them. This problem is less important now that systemic fungicides are available for controlling the causal organism.

Although soil maintained free of weeds chemically, without cultivation, provides a good firm surface for man and machines (Chapter 8), many growers prefer to retain a grassed strip between herbicide treated rows.

## REFERENCES


ATKINSON, D. (1977). Some observations on the root growth of young apple trees and their uptake of nutrients when grown in herbicidal strips in grassed orchards, *Plant and Soil*, **49**, 459–471

BUSHONG, J. W., GATES, D. W. and SULLIVAN, T. P. (1976). Mefluidide – a new concept in weed control with a plant growth regulator, *Proceedings 1976 British Crop Protection Conference – Weeds*, pp. 695–698

CLAY, D. V. (1980). The use of separate root and shoot tests in the screening of herbicides for strawberries, *Weed Res.*, **20**, 97–102

DAVISON, J. G. and BAILEY, J. A. (1979). Black polythene for weed control in young fruit and other perennial crops, *ARC Research Review*, British Growers Look Ahead Issue, pp. 11–14 (also see Parfitt *et al.*, Chapter 11)

FRYER, J. D. and MAKEPEACE, R. J. (eds.) (1978). *Weed Control Handbook*, vol. 2, *Recommendations*, Blackwell, Oxford, 532 pp.

KASASIAN, L. (1971). *Weed Control in the Tropics*, Leonard Hill, London, 307 pp.

LAWSON, H. M. and WISEMAN, J. S. (1979). Effects of raspberry suckers, growing in the alleys between rows, on cane and fruit production in a non-cultivated plantation, *Hort. Res.*, **19**, 63–74

MAFF (1977). *Chemical Weed Control in Bush and Cane Fruit*, ADAS leaflet STL 14, Ministry of Agriculture, Fisheries and Food, London, 18 pp.

MAFF (1979). *Chemical Weed Control in Strawberries*, ADAS leaflet HSF21, Ministry of Agriculture, Fisheries and Food, London, 26 pp.

NICKELL, L. G. (1976). Chemical growth regulation in sugar cane, *Outl. Agric.*, **9** (2), 57–61

RAHMAN, F. (1975). Weed Control in tea, *Outl. Agric.*, **8**, 173–177

ROBINSON, D. W. (1976). Orchard soil management, in *Fundamentals of Intensive Apple Production*, Handbook Series 8, An Foras Taluntais, Dublin, 219 pp.

ROQUES, J. F. (1976). Weed control in French vineyards, *Outl. Agric.*, **9** (1), 30–34

SETH, A. K. (1977). Integrated weed control in tropical plantations, in *Integrated Control of Weeds* (eds. J. D. Fryer and S. Matsunaka), University of Tokyo Press, Tokyo, pp. 69–88

STOTT, K. (1972). The effects of maleic hydrazide and 2,4-D for sward control in an orchard of Cox's Orange Pippin: 1965–1971, *Proceedings 11th British Weed Control Conference*, pp. 348–355

WATERHOUSE, D. P. (1980). Growth regulators in sward management, *Proceedings of Conference on Weed Control in Amenity Plantings*, University of Bath, 1980.

# 11 WEED CONTROL IN TURF, ORNAMENTALS AND FOREST TREES

## 11.1 MAINTENANCE OF LAWNS AND URBAN TURF

Grass sward fulfils many roles as a ground cover in gardens, on open spaces in town and country, on sports fields and beside roads, motorways and railways. Turf provides a good walking surface, is easily established from seed or by turfing and extensive areas can be maintained relatively cheaply by grazing or mowing. The species composition and method of management can be varied to produce a turf suitable for many different purposes. Because so many types of grass surface are required, the concept of weeds in turf is not easily defined. For example, small broad-leaved plants in flower add interest to grass in large open spaces, whereas the same species would be unacceptable in bowling greens or in small formal gardens. In some open spaces the greater part of the area may be encouraged to grow as an attractive meadow of grasses and broad-leaved species, public access being by paths that are more frequently mown. Fast growing, vigorous grasses such as Perennial Rye-grass (*Lolium perenne*) that recover quickly from damage are desirable on football pitches but would be out of place in fine lawns. The weed status of some annual grasses, particularly Annual Meadow-grass (*Poa annua*), is more difficult to define as these species often provide the only grass cover in poorly maintained turf.

The requirements of grass playing surfaces have been widely researched and chemical control of broad-leaved weeds is standard practice. Less attention has been paid to the labour intensive management of grassed areas in towns, such as those on grass verges, banks and between gravestones in cemeteries. The reduction in yield of crops caused by weeds is measurable in terms of profit and loss, whereas the management of small urban grassland areas can only be regarded as satisfactory when its appearance is acceptable to the community. The requirement is essentially for tidiness throughout the season, but public authorities are under increasing pressure to reduce costs.

Recently, the public has come to expect turf to be kept neat and tidy by regular and frequent mowing. If the frequency of cutting is reduced in an attempt to save labour then extra time must be spent picking up and removing the grass, so that the net saving is nil (*Figure 11.1*). An examination of the factors that make urban turf

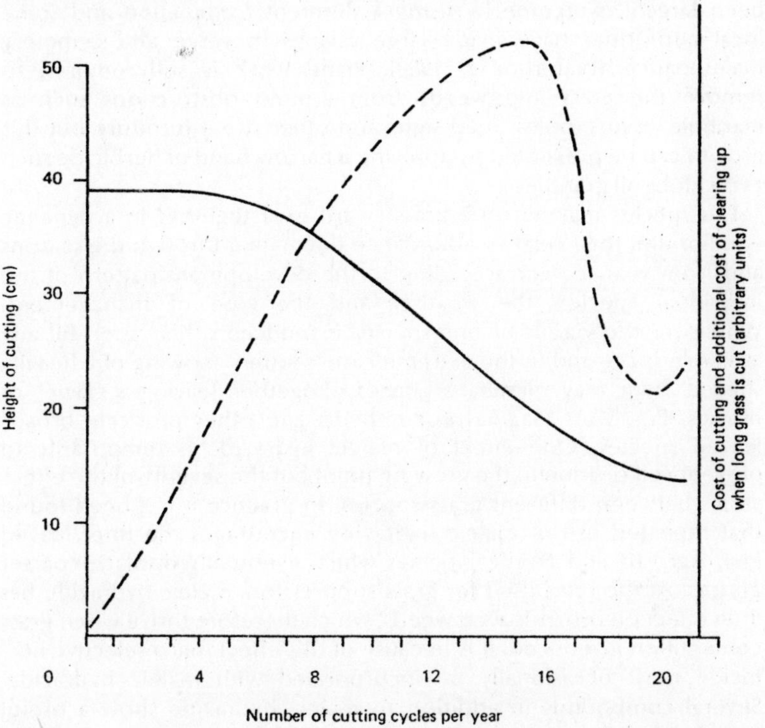


**Figure 11.1** Comparative costs of cutting grass on small grass areas in Plymouth using pedestrian controlled mowers, including costs of clearing up longer grass when necessary. Solid line indicates the grass height at time of cutting according to mowing frequency. Broken line indicates the relative costs of cutting and any necessary clearing up according to mowing frequency. (From data supplied by D. P. Waterhouse.)

visually acceptable suggests that evenness of height and colour are more important than absolute height and that low growing broad-leaved plants without flowering stems are probably less offensive than tall flowering grasses. Because urban grassed areas are frequently interrupted by driveways, poles and other structures mowing is always highly labour intensive. Efforts are being made to achieve an acceptably tidy appearance at a greatly reduced cost by spraying with a combined herbicide and growth regulator. The herbicide, usually MCPA or mecoprop, controls broad-leaved weeds, while the growth regulator, maleic hydrazide, suppresses grass growth sufficiently to allow a much reduced mowing programme. The problems of scorch that have been encountered with maleic hydrazide appear to have been largely overcome by using a different formulation and some local authorities have made large savings in verge and cemetery maintenance (Waterhouse, 1980). Hand work is still required to remove the grass and weeds from around obstructions such as manhole covers, poles, road signs and other street furniture but this growth can be prevented by applying a narrow band of herbicide such as dichlobenil granules.

The species making up a grass sward exist together in a dynamic equilibrium, their relative abundance fluctuating through the seasons and from year to year according to the development pattern of the individual species, the weather and the type of management. Whenever the vigour of one species is reduced others soon fill any available space and in the extreme case frequent mowing of a heavily shaded lawn may eliminate grasses altogether leaving a 'lawn' of mosses, Pearlwort (*Sagina procumbens*) and other prostrate broad-leaved species. One effect of maleic hydrazide is temporarily to prevent cell division in the growing point but the severity of the effect varies between different grass species. In practice it has been found that repeated use of maleic hydrazide encourages the finer leaved *Poa*, *Agrostis* and *Festuca* species which eventually displace coarser grasses. At the rates used for grass suppression, maleic hydrazide has little effect on broad-leaved weeds, which therefore thrive when grass competition is reduced. It is because of this effect that a selective herbicide must occasionally be incorporated with maleic hydrazide. Several compounds in addition to maleic hydrazide show a useful retardation of grass growth (Shearing and Batch, 1979), including mefluidide which is commercially available for this purpose in North America.

### 11.1.1 Weed problems in turf

Regularly mown turf provides a hostile environment for many species of plants and few weeds of cultivation can survive in turf. However, newly sown lawns suffer as much from weed competition as do young crops and where the cost is justified prior soil sterilisation with dazomet or methyl bromide will provide the grass with a weed-free seedbed. Alternatively a period of fallow before sowing will allow germinating annual weeds and the more persistent perennial ones to be destroyed with appropriate herbicides; in practice this technique is most suitable for autumn sown grass. Grass seedlings of the finer leaved lawn grasses are easily damaged by the herbicides based on 2,4-D, MCPA or mecoprop that can safely be used later on more mature grass swards but ioxynil can be applied safely to most grass seedlings provided that they have at least two fully expanded leaves. Ioxynil will control many common broad-leaved weeds at the seedling stage but Knotgrass (*Polygonum aviculare*) and White Clover (*Trifolium repens*) are scorched but not killed.

A specialised flora of weed grasses and broad-leaved species compete effectively with the more desirable turf grasses when they are allowed to become established. Several weeds, including Slender Speedwell (*Veronica filiformis*) can rapidly dominate lawns once introduced. This weed produces no seed but small stem cuttings cut and scattered during mowing root at the nodes providing effective propagation and dissemination. Several other invasive perennials are common in lawns, including Yarrow (*Achillea millefolium*), several other Speedwells (*Veronica* spp.), common Mouse-ear (*Cerastium holosteoides*) and Selfheal (*Prunella vulgaris*); in addition the rosette-forming species, including Dandelion (*Taraxacum officinale*), Daisy (*Bellis perennis*) and Greater Plantain (*Plantago major*) compete by shading the grass under overlapping ranks of leaves. The rosette formers are the least desirable in almost all types of sward, especially the plantains that die back in autumn leaving dead leaves and a bare patch. All broad-leaved lawn weeds can be controlled chemically, although the more persistent ones require repeated application; but their removal with selective herbicides may increase problems due to undesirable or 'weed' grasses. Annual Meadow-grass (*Poa annua*), for example, which is an undoubted weed in fine turf, cannot be controlled selectively without destroying other grasses. This species, which behaves in turf as a short lived perennial rather than as an annual, is extremely variable in form. In many situations, including sports fields, *P. annua* colonises bare areas and provides a valuable green cover at times when cover from other grasses is inadequate. In parts of North America lawns composed mainly of Smooth Meadow-grass (*Poa pratensis*), known locally as Kentucky Blue-grass, and bent grasses (*Agrostis* spp.) become prone to invasion by annual kinds of Crab-grass (*Digitaria* spp.) which can germinate and grow in the hot continental summer at the expense of the lawn grasses; these Crab-grasses are considered in North America to be the worst weeds in turf but fortunately they can be controlled by pre-emergence application to the bare patches of soil-acting herbicides such as chlorthal-dimethyl or oxadiazon. In Britain the soft light green of Yorkshire-fog (*Holcus lanatus*) and Creeping Soft-grass (*H. mollis*) spoil the uniform dark green of mown turf but at present only physical removal can be recommended. Other grasses that can be considered as weeds include the strong growing and drought-resistant perennial Cocksfoot

(*Dactylis glomerata*), producing dense tufts of coarse grass which persist despite mowing, and the annual Wall Barley (*Hordeum murinum*), thriving on disturbed sites and where turf abuts on to pavements and walls and producing unsightly flower spikes in summer and autumn.

Legumes, such as White Clover (*Trifolium repens*) and several small trefoils able to withstand mowing, may become abundant when soil phosphorus is adequate and nitrogen scarce because they possess bacteria containing root nodules that fix atmospheric nitrogen. Clovers are usually regarded as weeds in formal and sports turf, but their status elsewhere depends on personal judgement. They can be discouraged by adding nitrogen fertiliser and can be controlled selectively in grass with mecoprop; the encouragement of clovers at the expense of grass can be achieved by applying propyzamide (Chapter 10). The great value of White Clover and other legumes is during the establishment and maintenance of grass sward on derelict or degraded land such as mineral spoil heaps or on the disturbed sites largely lacking in top soil that follow construction work. The main reason for poor grass growth on these soils is lack of nitrogen, and much the best way of supplying it is by using a legume as the pioneer species and later in the sward also (Johnson and Bradshaw, 1979). Clovers can fix over 100 kg N/ha/year, and because the organic nitrogen produced in the nodules is easily mineralised it is soon made available to the grass roots.

Although moss gardens are much prized by the Japanese, and fruit growers have come to appreciate the benefits of a moss ground cover beneath herbicide treated fruit bushes, excessive moss growth in lawns generally indicates poor growing conditions for grass. Factors that reduce the vigour of grass and encourage moss include heavy shade, lack of drainage, excessive soil acidity (pH below 5.5), insufficient aeration following repeated compaction and a mat of dead and decaying plant remains at the soil surface. Some or all of these are commonly found in poorly maintained lawns of private gardens. High spots, which are repeatedly scalped when mowing, are invaded by moss and other prostrate weeds. There are distinct moss types, some appearing in dry situations, others growing better in constantly damp ground, but their appearance always indicates the inability of the grass to maintain a dense, tight sward. Although mosses in lawns are quite easy to kill chemically with chloroxuron or lawn sands containing ammonium sulphate and ferrous sulphate, their temporary destruction does nothing to cure the underlying causes of the moss problem.

The need for good management applies equally to the prevention of other weed problems. Despite the use of good quality grass seed largely free from weed contamination, weeds become established from imported seed or weed seed already in the soil. In rural areas

Dandelion (*Taraxacum officinale*) is frequently a successful invader of lawns after seeds are blown in from nearby grassland. In addition, seeds produced *in situ*, especially those of *Poa annua*, are easily spread when mowing. Another source of weeds are the seeds brought up to the surface in worm casts.

The finer species of *Agrostis* and *Festuca* grow well between pH 5.5 and 6.0 which is rather too acid for most lawn weeds, so unless the soil is already sufficiently acid the use of acid fertilisers such as ammonium sulphate will be beneficial; however, excessive acidity, which favours a specialised non-grass flora including some mosses, should be avoided. Whatever is done to encourage a vigorous, healthy sward will discourage invasion by weeds.

## 11.2 MAINTENANCE OF ROUGH GRASS AREAS

The principal objective in the management of countryside parks, rural roadside and motorway verges, railway embankments and other areas of rough grass is to establish a plant cover that is visually attractive, blends with its surroundings and protects the surface from erosion. These objectives must be achieved without creating either a fire risk or a source for the dissemination of noxious weeds, and on roadside verges without obstructing signs or sight lines. In the past, rough grass areas have been maintained by regular cutting. By controlling the time of cutting it is possible to prevent undesirable species such as Spear Thistle (*Cirsium vulgare*) from increasing or spreading by seed. Cutting was often done by hand which made a form of selective control of the less desirable species possible; this is no longer available when the task has been mechanised.

### 11.2.1 Noxious and other troublesome weeds in grass

In Britain the Weeds Act, 1959, requires landowners to prevent certain noxious weeds from spreading their seeds to adjoining property. These weeds are Creeping Thistle (*Cirsium arvense*), Spear Thistle (*C. vulgare*), Broad-leaved Dock (*Rumex obtusifolius*), Curled Dock (*R. crispus*) and Common Ragwort (*Senecio jacobaea*). Appropriate legislation in other countries specifies named noxious weeds, requiring their destruction or at least that their presence be reported. In addition to the weeds named in the Weeds Act there are many more that establish quickly on disturbed ground such as motorway verges on which grass will be sown, including Colt's-foot (*Tussilago farfara*), Rosebay Willowherb (*Epilobium angustifolium*), other willowherbs and Dandelion (*Taraxacum officinale*); like those listed in the Weeds Act all can be wind blown on to farm land. Weeds may be objectionable for other reasons, such as the fast growing Giant Hogweed (*Heracleum mantegazzianum*) and Giant Knotweed (*Polygonum sachalinense*), both of which grow above 2 m in less than one year. A similar problem is posed in North America by Giant Ragweed (*Ambrosia trifida*), a notorious cause of hay fever, which often grows alongside roads to a height of more than 4 m, obscuring road signs and sight lines.

Introduced weeds sometimes spread from their initial infestation on roadsides and railway banks to nearby farm land. Most such introductions are accidental as was that of *Parthenium hysterophorus*, taken to Queensland in the 1950s (Chapter 2), probably with imported cattle feeding stuff. In other cases it is deliberate, as it was when Kudzu (*Pueraria thunbergiana*), a scrambling legume first introduced from Japan, was planted in the southern parts of North America to protect steep road embankments from erosion. To the dismay of local farmers and foresters the fast growing Kudzu vine soon invaded adjacent farm land and forest, smothering all in its path, even tall forest trees.

### 11.2.2 Motorway verges

Maintenance of existing grassed areas must be done at the lowest cost and the frequent mowing that was previously the norm for motorway verges is no longer permitted. In a technical memorandum sent to local authorities in Britain in 1975 the Department of the Environment stated that 'in future grass on land forming part of trunk roads and motorways is to be cut only in certain restricted places and circumstances.' The exceptions to this general rule are on grounds of safety or of amenity in built up areas. Some European countries still allocate resources to keeping motorway and roadside verges 'neat and tidy' but in Britain the abandonment of mowing on all but a 2 m strip beside the hard shoulder is now seen as a virtue since it has allowed motorways to blend into their rural setting, their formerly clearly defined edges now blurred.

To change from existing mown turf to a rougher and taller vegetation, in which coarse grasses, perennial 'weeds' and perhaps some scrub plants grow freely, does not present too great a challenge in the short term, although future management as the scrub matures and spreads will be more difficult. A much greater challenge is the low cost management of new and often poorly prepared sites in a way that avoids them becoming an unsightly wilderness of tall, weedy perennials. The phase of undesirable perennial weeds has been prevented by sowing a fast growing annual grass, Westerwolds ryegrass, which prevents the growth of the taller, noxious weeds while allowing native grassland species to take over as the ryegrass dies (Wells, 1980). Careful planning and land preparation at this stage would also simplify subsequent maintenance. More research is required into the control of the natural succession on disturbed sites such as motorway verges and similar newly created habitats by management methods.

### 11.2.3 Chemical manipulation of rough grass

A potential management tool now being assessed is the chemical manipulation of rough, unmown grass sward to encourage 'desirable' species and discourage others. In urban grassed areas maleic hydrazide has been successful in reducing the frequency of cutting required while in North America mefluidide is used for suppressing grass sward beside highways; however, both compounds are probably too expensive to use where mowing with large machines is practicable. In addition, there are several herbicides that have the potential for changing the grass species composition of sward, such as ethofumesate, which favours ryegrasses at the expense of unproductive 'weed' grasses in leys used for cattle forage. Similarly, paraquat-resistant perennial ryegrass can be grown as a pure stand by destroying other species with paraquat. Glyphosate could perhaps be used to destroy all plants that grow above a predetermined height using a simple 'wet rope' applicator (Dale, 1979; Lutman, 1980). There is much potential for a simple and inexpensive chemical manipulation of extensive amenity areas to produce swards suitable for particular purposes.

## 11.3 PRODUCTION AND MAINTENANCE OF WOODLAND AND FOREST TREES

There is a great variety of tree planting situations in Britain, including large tracts of coniferous forest for timber production, smaller areas of specialist production such as beds of willows for basket making and smaller scale plantings of windbreaks, visual screens and specimen trees. The possibility of short-term coppicing of fast growing species such as willows or poplars as a renewable source of energy and chemical industry feedstock is now being investigated. Should such schemes prove workable and economically competitive with other fuel sources then effective weed control will be essential to maximise fuel yields.

Woodland and forest planting may be in land new to forest, in the lowlands generally former agricultural land and in the uplands existing open moorland; increasingly, however, the forester is re-establishing forest on land cleared after the 40-year cycle common in Britain. Agricultural land is already rich in weed seeds and replanted areas build up weed populations rapidly from seeds that either survive the 40-year cycle or are blown in from outside. On new and relatively infertile sites many of the species present are well adapted to open woodland conditions. Thus young trees in woods and forests are always presented with a competitive weed flora of annuals and perennials arising from seed; problems are also caused by deep-rooted and aggressive perennials including Heather (*Calluna vulgaris*), Bracken

(*Pteridium aquilinum*) and Rhododendron (*Rhododendron ponticum*) that survive pre-planting preparation.

### 11.3.1 Tree seedbeds

Most forest trees are raised from seeds in nurseries, where they remain for 12–18 months. During the whole of this time the slow growing and poorly competitive trees must be protected from weeds. Traditionally this has been done by hand, workers often needing as much as 10 minutes to remove weeds from one square metre of bed; regular hand weeding of an estimated 450 hectares of forest nursery in Britain is obviously a huge task. Paraquat applied pre-emergence, particularly when a stale seedbed is used, will delay the need for the first weeding, but greater savings can now be made by overall sprays of diphenamid, a soil-acting herbicide that controls annual grasses and some seedlings of broad-leaved weeds. An application before the tree seedlings emerge, followed by up to three more during the lifetime of the nursery fails to control all weed growth but the reduction in weediness is so substantial that up to 75 per cent of weeding costs may be saved (Tait *et al.*, 1980). Diphenamid is tolerated by most commonly grown conifers and broad-leaved trees but Japanese larches and birch are sensitive. Nitrofern is claimed to be almost as effective but more frequent applications are required. Other herbicides are under test for nurseries, some of which may have advantages for certain tree species, but none has been found to combine the safety and effectiveness of diphenamid (Biggin, 1979).

Many kinds of tree seedlings grow as well (as judged by increase in height or dry weight) under partial shade as in full sunlight. Helliwell and Harrison (1979) found little reduction in growth of sycamore, ash, birch or larch but a halving of weed dry matter increase when the daylight was reduced to 43 per cent of its full value. However, it is probably cheaper to use herbicides than to erect expensive shading over tree nurseries.

The methods described for raising forest trees are reasonably cheap but do not always produce young plants with the ability to cope with adverse conditions when planted out. A higher percentage of success has been claimed for plants raised in small paper pots containing sterilised soil, similar to those used in Japan for sugar beet, that can be planted by machine.

### 11.3.2 Transplant lines

From the nursery seedlings are generally moved to temporary 'transplant lines' to free them from a build up of resistant weeds and to encourage compact rooting (Biggin, 1980). Simazine is a safe herbicide for many tree species in transplant lines but European larch, *Picea omorika* and ash, amongst others, may be damaged; atrazine may also be used subject to some limitations (Fryer and Makepeace, 1978). Annual grasses (mostly Annual Meadow-grass, *Poa annua*) and perennial grasses are often the dominant weeds and effective selective control among young trees can be obtained with propyzamide, applied either as granules or a spray during the winter months. This compound must be absorbed from the soil; it then acts herbicidally by inhibiting growth of weed roots. Because breakdown in moist soil above 5°C is relatively rapid, application other than in the winter is less effective against the rhizomatous grasses such as Common Couch. Perennial broad-leaved weeds such as Sheep's Sorrel (*Rumex acetosella*) are not well controlled by propyzamide but both glyphosate and hexazinone, which are more effective, can be used selectively in conifers subject to certain limitations.

### 11.3.3 Established tree plantings

Herbicides that have been recommended for forestry include asulam, atrazine, dalapon, dichlobenil (and the similar chlorthiamid), 2,4-D, fosamine ammonium, glyphosate, hexazinone, propyzamide, 2,4,5-T and trichlopyr; individual compounds or mixtures have been proposed for particular problems, subject to certain limitations. Problems in application, often on steep or uneven ground and in inaccessible places, have encouraged the use of low-volume spraying from aircraft or helicopter, and of hand- or tractor-mounted CDA equipment using spinning discs; granules have also been adopted for some herbicides.

### 11.3.4 Forest weeds

Woody weeds often thrive in the absence of competition following herbicide spraying, and heathers (*Calluna vulgaris* and *Erica* spp.) can have a debilitating effect on trees, especially on spruces, when planted on poor acid soils. This stunting can be so severe that when *Calluna vulgaris* is allowed to regenerate after tree planting and encroach on young Sitka spruce, extension growth may cease – the spruce is said to be 'in check'. This ability of the heather has been claimed to be due to allelopathic substances released into the soil by the weed. However, competition may result from the inability of the mycorrhizal roots of a slow growing conifer to obtain sufficient phosphorus and nitrogen from an impoverished acid soil in the presence of the mycorrhizal roots from vigorous heather rather than from an allelopathic interaction. The importance of nutrition is suggested by the alleviation of the checked condition when spruce trees are interplanted with Irish Furze (*Ulex gallica*) or with Japanese Larch; the former fixes nitrogen whereas the latter returns nutrients in fallen needles, but in either case nitrogen becomes available to the spruce. Alternatively, the trees can be taken out of check by chemical removal of the heather with 2,4-D.

Another woody weed problem in Britain is Rhododendron (*Rhododendron ponticum*), an aggressive alien weed that is poisonous to livestock and also indirectly to man, through honey made from its blossom. Rhododendron, which forms dense thickets and can grow to 5 m, has become naturalised, often invading and suppressing both native woodland and planted forest. Following deliberate plantings in Ireland in the 1920s the weed has spread at dramatic speed into adjacent plantations. Trees set seed when about 12 years old, so that after felling and clearance of an infested forest the plant regenerates and infests the new plantation, both from seed and from basal buds on the remains of the old plant (Robinson, 1980). Herbicides provide control and subject to some limitations 2,4,5-T, trichlopyr and glyphosate can all be used; if seed has been allowed to form new infestations will arise and repeated treatment will be necessary.

### 11.3.5 The problem of replacing 2,4,5-T in forestry

The herbicidal properties of 2,4,5-T, which has been in routine use in forests for controlling woody weeds for over 30 years, are known and predictable. Its relatively low acute toxicity to mammals and its wide spectrum of selective activity within coniferous forest have made it popular for 'conifer release' from competing broad-leaved trees. It has also been the main herbicide used in North America for controlling woody plants that encroach on 'management rights of way' beside electric power lines and pipelines.

It is believed that 2,4,5-T containing less than 0.1 ppm of TCDD (dioxin), which is now the maximum allowed in most countries, does not present an unacceptable hazard as a herbicide (MAFF, 1979). Nevertheless public concern over its safety has led some governments to restrict its use (Chapter 6). Several alternatives, including fosamine ammonium, glyphosate and triclopyr, have been considered, and in Norway control of broad-leaved woody weeds in forests is now achieved by carefully timed applications of glyphosphate by ground sprayers or from helicopters.

## 11.4 AMENITY TREE AND SHRUB PRODUCTION

The debilitating effects of weed competition on young trees and bushes in the nursery was shown by Davison and Bailey (1980) who grew five species of nursery stock plants chosen to represent a wide range of types, and allowed the weeds to grow unchecked for varying periods during two growing seasons. The species were *Acer platanoides, Chamaecyparis lawsoniana, Philadelphus* x *virginalis, Potentilla fruticosa* and *Prunus laurocerasus*. They were all reduced in size when weeding was delayed beyond early June or in some cases late May in the first year, and even in the second year well grown plants of *Potentilla, Prunus* and *Chamaecyparis* which had been weed free in the first year were stunted by the presence of weeds, but presumably the remaining three species were sufficiently vigorous at this stage to compete effectively with the weeds. The better quality of the plants produced in the absence of weeds represented a significant increase in the value of the plant crop. This investigation also showed the value of a black polythene mulch over and above that due to its control of weeds, producing the best plants of any treatment; this effect of black polythene is probably due to improved water conservation. If the growth of woody plants is regarded as a form of compound interest then reduced growth in any one year is responsible for increasingly large deficits in subsequent years. Fruit growers appreciate the need to prevent weed growth around their plants so as to ensure rapid unhindered development and a rapid return on their investment.

The problems of nursery stock production and its weed-free maintenance on site are very similar to those met in bush and tree fruit production and the same soil- and foliar-acting herbicides are used (Chapter 10). Information on the susceptibility of woody ornamentals to herbicides has been accumulated slowly but the testing has not been done with the same thoroughness as with fruit and plantation crops. However, sufficient is known to allow routine use of herbicides in the production of nursery stock, both in containers and in the open ground.

### 11.4.1 Herbicides for nursery stock

The choice of herbicides depends on the susceptibility of the species and cultivars grown. Tolerance may vary on different soil types, at different times during the growth cycle and with the age of the plant; plants are considered at greater risk soon after planting when some roots may be exposed. Control of particular weeds, especially of perennial species, may require application at precise growth stages or times of the year for maximum effect. Paraquat will destroy the aerial parts of most weeds and effectively controls annual weeds, but deep rooted perennials soon recover; however, Creeping Buttercup (*Ranunculus repens*) can easily be destroyed with paraquat which is safe to apply in the winter provided the spray does not reach green bark or leaves. Soil-acting herbicides have been recommended with certain reservations (Fryer and Makepeace, 1978) for seedling weed control in many types of nursery stock. Perennial weeds are more difficult to control and those not destroyed by pre-planting treatment can be spot treated with an appropriate herbicide or mixture (*Table 11.1*).

*Table 11.1*

*Herbicides that have been suggested for nursery stock*

| | |
|---|---|
| *Overall treatment* | *Mainly for weed seedling control* |
| Alloxydim-sodium | Annual and perennial grass control |
| Dichlobenil | Can cause damage – recommendations must be followed precisely |
| Lenacil | Kills some simazine-resistant weeds |
| Oxadiazon | Apply early spring before growth begins |
| Paraquat | For emerged weeds; no residual action |
| Propachlor | Applied as spray or granules |
| Simazine | Generally safe but some species sensitive |
| Trifluralin | Soil incorporated before planting |
| *Spot treatment* | *Mainly for perennial weed control* |
| Aminotriazole | Apply only among well-established plants |
| Asulam | Effective for dock control |
| Dichlobenil | Spot treatment easy with small granule spreaders |
| Glyphosate | Very effective against perennial grasses and some broad-leaves. Can be applied as spray or in alginate jelly |
| Oxadiazon | Effective against bindweeds (*Convolvulus* and *Calystegia* spp.) |
| Paraquat | Destroys Creeping Buttercup (*Ranunculus repens*) during winter |
| Propyzamide | Effective against perennial grasses applied during winter |

### 11.4.2 Nursery stock plants in containers

The production of woody plants in containers has created some unusual weed problems. Beginning with sterilised soil, containers are initially free from weeds but the fertile, well-watered soil provides an ideal environment for the germination of imported seeds. These include both annual and perennial Willowherbs (*Epilobium* spp.), Dandelion (*Taraxacum officinale*), Colt's-foot (*Tussilago farfara*) and the seeds of birch, willow, sycamore, buddleia and pines. Ephemeral weeds such as Annual Meadow-grass (*Poa annua*) and Groundsel (*Senecio vulgaris*), which produce several generations per year, can build up rapidly from a few pioneer plants, but the most notorious weed is Hairy Bitter-cress (*Cardamine hirsuta*), whose explosive seed pods can spread the weed rapidly throughout the nursery. Liverworts (mainly *Marchantia* spp.) and mosses thrive in the fertile, often damp conditions and must be controlled but they are unsightly rather than competitive.

Introduction of imported seed can be reduced by careful timing of the containerising or 'canning' operation in relation to the release of windborne seeds and by keeping the sterile ingredients of the container soil covered until ready for use; problem weeds such as Hairy Bitter-cress must be prevented from establishing by prompt removal of pioneer plants. But weed problems will arise despite every precaution and herbicides then provide the only alternative to costly hand weeding. Most weeds from seed can be prevented by the use of soil-acting herbicides that destroy the young seedlings, but because of the restricted root space in containers compared with plants in the open ground, the choice of safe herbicides is more limited. Overall treatment with chloroxuron at a high spray volume has been found to be safe on most species but 'soft' growth can be scorched, particularly when under glass or polythene protection, and a thorough washing with water after application is recommended. Repeated applications of chloroxuron are required and annual grasses are not well controlled so that longer lasting and more effective materials including simazine and lenacil have been used safely for some species (Fryer and Makepeace, 1978). Oxadiazon has also been recommended for woody container plants in North America but as this herbicide can be absorbed through leaves as well as through the soil the spray must be kept off the green aerial parts. Application by granules, available in some countries, therefore has advantages. This herbicide has a wide weed spectrum, including several simazine resistant species, but it fails to control common Chickweed (*Stellaria media*).

## 11.5 AMENITY TREE AND SHRUB PLANTINGS

Weed competition must be prevented both in the nursery and during early establishment in its permanent site, so as to keep woody ornamentals in an actively growing condition. Rapid growth is no longer required when individual specimen plants become well established and competition from grass sward need no longer be prevented. This would also be true of tree screens planted to hide ugly and intrusive structures or views, in which initially rapid growth in weed-free surroundings can be followed by slower growth in grass when the trees have reached the desired height and density.

In amenity woodlands, where wild plants and animals may be viewed by the public from nature trails, the use of chemical weed control would be undesirable except at the earliest stage of plant establishment or later to control by spot application species considered undesirable. This can be contrasted with formal urban amenity plantings in which weeds, while perhaps of negligible competitive significance, may be unsightly and therefore demand control.

### 11.5.1 Chemical weed control

Chemical weed control has become an integral part of the practice of nursery stock production, but there remains suspicion and fear of herbicide use in permanent plantings. This low acceptance is due in part to ignorance and misunderstanding of the best ways to utilise chemical weed control and a mistrust of its safety, but the difficulty of applying herbicides accurately to small and irregularly shaped areas, which may themselves be surrounded by other and more susceptible plants, must also be a factor; in addition, some people are concerned when they see herbicides being applied in a public place, imagining that because the operators are wearing face shields and other protective clothing that the chemical being applied must therefore be poisonous or dangerous. The public require reassurance that chemicals being used in parks and other open spaces are entirely safe to them, their children and to pets.

Herbicides must be used with caution immediately after planting ornamental woody plants in their permanent site because the soil may not have settled around the stem and with careless planting some roots may be exposed. Some woody plants possess an inherent tolerance of simazine, although not wholly resistant. Others show little tolerance but depend for their safety on the herbicide remaining very near the soil surface (Chapter 6). Robinson and Kelly (1976) have shown that species that regenerate easily from seed despite regular simazine application, including *Cupressus macrocarpa, Erica arborea, Berberis darwinii, Cotoneaster frigidus* and *C. simonsii*, are tolerant of simazine present in the rooting zone. Plants that do not regenerate under these conditions, including species of *Betula* and *Eucalyptus*, were found to be susceptible. The latter group are more likely to be damaged when application follows careless planting that allows access to the roots. Contact herbicides, such as paraquat, and soil-acting ones including simazine and lenacil have been recommended (Fryer and Makepeace, 1978). Sprays may be applied with a sprayer such as the 'Arbogard' which delivers a pre-determined dose of herbicide to an area immediately around the plant stem.

### 11.5.2 Use of black polythene mulch

An alternative method of preventing weed growth is to fix black polythene sheet in strips or patches to the soil surface immediately around the stems of individual trees or shrubs, leaving it in place until the leaf canopy effectively discourages further weed growth (Parfitt *et al.*, 1980). At the simplest, used polythene fertiliser sacks can be secured with soil at each corner, a slit having been cut for the tree stem; where appearance or longevity are important the sheet can be hidden, and protected from ultra violet light, by a decorative covering of stones or other inert and sterile material that will not support weed growth. A 15 cm depth of sawdust or bark chippings is often used as a mulch in North America and continental Europe to discourage weeds around trees, but rodents that attack the living tree bark may have to be controlled. Small infestations of perennial weeds, including grasses, can generally be destroyed by spot treatment with glyphosate; this herbicide also has been used to control suckers in *Prunus* species without harming the parent plant.

It is obviously sensible when planning a mixed planting to choose species and cultivars possessing similar herbicide tolerance so that the whole planted area can be treated with the same herbicide or mixture (Robinson, 1975/76).

### 11.5.3 Woody ornamentals that become weeds

For ornamental plants to thrive in competition with other species they must themselves be vigorous. Species that produce many viable seeds and have an effective method of distribution, or possess an invasive root or shoot system have sometimes spread out of control and become troublesome weeds (Chapter 2; *Rhododendron ponticum*, this chapter). In Britain seedlings of Norway Maple (*Acer platanoides*), an introduced tree species, often become established after wind dissemination of the seeds; in the same way seedlings of *Cotoneaster horizontalis* and other *Cotoneaster* species and of *Leycesteria formosa* grow from seeds carried by birds. These plants often survive treatment with soil-applied herbicides and must be removed by hand or killed with an appropriate translocated herbicide. In North America there are many examples of such escapes and care should be taken to avoid the rampant spread of new introductions. *Grevillea chrysodendron* is an attractive ornamental tree from Queensland which appeared to be suitable for use in Florida for low maintenance planting but before release several plants were grown in isolation for assessment. After six years the plant's increase was so rapid that all the plants were destroyed because of its 'prodigious fecundity' and its proven ability to colonise fallow land in Florida (Neel and Will, 1978).

## 11.6 BULBS AND CORMS

Bulb productivity is reduced by weeds present during the vegetative phase of bulb development, when next year's bulb is being produced (Chapter 1); weeds also interfere with flower picking and bulb harvesting. Work on chemical weed control has been mainly devoted to narcissus, of which Britain is the world's largest producer, but most of the results are equally relevant to other kinds.

The long periods during which bulbs and corms of narcissus, tulip, dutch iris and gladiolus remain below the ground without foliage provides good opportunities for chemical weed control; in addition, certain herbicides can be applied post-emergence. Paraquat, often combined with one or more soil-acting residual herbicides, applied to weeds just before bulb emergence destroys existing weeds and provides prolonged control. Suitable residual herbicides include chlorpropham, chloridazon/chlorbufam formulated as a mixture, lenacil, linuron and simazine. Application of simazine should be restricted to heavier soils, in which it is unlikely to reach the bulbs. Several of these herbicides, but not simazine or paraquat, can also be applied post-emergence to destroy weeds that germinate later as the effects of the earlier application diminish (Fryer and Makepeace, 1978; MAFF, 1978). At the end of the period of flowering and vegetative growth when the bulbs or corms are fully senescent the bulbs may be lifted, but if a serious regrowth of weeds has occurred by this time a non-translocated contact herbicide such as dinoseb (in an oil formulation) or sulphuric acid will destroy it without harm to the bulbs. But if the bulbs are to remain for another year or more perennial weeds must be controlled at this stage also. Provided that all leaves have been separated from the bulb, herbicides such as aminotriazole, asulam or paraquat can be applied. Aminotriazole is effective against rhizomatous grasses and several broad-leaved species, asulam against docks (*Rumex* spp.) and paraquat will kill Creeping Buttercup (*Ranunculus repens*). Oxadiazon has been suggested for spring planted gladiolus, applied shortly after planting; but probably is less safe for autumn planted bulbs. Some control of weeds with chemicals is also possible for the many kinds of bulbs individually grown on a small scale but fewer firm recommendations have been made.

In the bulb fields of Southwest England and the Scilly Isles bulbous *Oxalis* (mainly *O. latifolia* and *O. pes caprae*, respectively) can cause severe problems. The small bulbils easily become detached and are spread with soil on boots, implements and planting material. The *Oxalis* bulbs can be destroyed by applying pre-planting chemical sterilants and control can be achieved also by grassing down for several years, but the former is expensive and the latter often impracticable on small intensive bulb farms. However, some control is achieved by spraying trifluralin and thoroughly incorporating it throughout the infested soil.

## 11.7 OTHER HERBACEOUS PERENNIALS

A great variety of hardy and half-hardy herbaceous perennials are grown, both for cut flowers and in planting schemes. Most work has been done on those grown for cut flower production, such as chrysanthemums and dahlias. For chrysanthemums a pre-planting soil-incorporated spray of trifluralin can be followed by chloroxuron applied overall after planting, or pentanochlor (a contact herbicide with some residual action) as a directed spray to avoid green leaves and young stems. For dahlias both lenacil as an overall spray and propachlor granules can be applied for controlling germinating weeds.

Many of the other kinds of herbaceous perennials grown in gardens and planting schemes can tolerate one or more of the herbicides mentioned for chrysanthemum and dahlia. Lenacil, for example, which controls the seedlings of a wide range of weeds, is widely tolerated by established plants. Propachlor granules are also recommended for herbaceous plantings, including annual bedding plants, but again only germinating weeds are controlled. Perennial weeds present a more difficult problem but application of glyphosate either as a coarse directed spray or made up as a gel and painted on the leaves can be effective; when a gel is used care must be taken to avoid excess gel dripping on to the leaves of ornamental plants. Where the weeds are significantly taller than the ornamental plant then rope applicators, which have proved safe and effective in some crops, could probably be used. Another method worth trying when perennial weeds invade is to apply glyphosate directly to the weeds with the herbicide glove (Chapter 7).

Both alloxydim-sodium and fluazifop-butyl may be recommended for controlling perennial grasses in some herbaceous perennials.

## 11.8 TOTAL WEED CONTROL

There are many situations in which all weed growth must be prevented, either permanently or temporarily. The immediate surroundings of oil refineries, chemical factories and stores, and railways and airfield runways must be free from weeds for safety reasons. On a smaller scale pathways, car parks and the central reservations of urban motorways also require total weed control. When chemicals are used, especially on relatively non-absorbent surfaces such as concrete or asphalt, there is a risk of run-off of damaging quantities of herbicide reaching specimen trees within the treated area or any plants adjacent to it. When porous pathways are constructed it is good practice to use a self-binding gravel or 'hoggin' which contains clay beneath the top surfacing gravel; this clay fraction helps to prevent the lateral spread of herbicide to nearby areas. Factors that must be considered if damage is to be avoided include the rainfall, the kind of surface, the slope of the ground and the type of herbicide. If the treated land is later to be used for planting then the rate of loss by breakdown or leaching may also be important. Alternative methods of controlling weeds, such as burning or cultivating, may sometimes have advantages but the material and labour costs of spraying herbicide are usually less than those of other methods.

There is a wide choice of suitable herbicides and mixtures for total weed control and because crop safety can generally be disregarded higher rates of herbicide providing more prolonged weed control can be applied. Primarily foliar-acting materials such as aminotriazole, paraquat, glyphosate and 2,4,5-T or 2,4-D are often combined with mainly soil-acting ones including atrazine, borates, bromacil, diuron, monuron and simazine, but only certain mixtures are satisfactory; glyphosate should not generally be mixed with other herbicides. Granular herbicides including atrazine, dichlobenil and oxadiazon are suitable for spot treatment. Sodium chlorate, which is foliar and soil acting, is an effective herbicide but its effects may be relatively short lived and it has been largely displaced by more modern materials. Picloram, which also acts through leaves and soil is particularly effective against some of the deep rooted perennials that survive other treatment. (This herbicide is available in Britain for total weed control, but not for use in crops.) Herbicides and mixtures will generally be used in a rotation to prevent the build up of particular weed species.

Different sites require a variety of application methods, but traditional high-volume spraying is the more usual practice. Where access is difficult or where speed of application is essential then low-volume spraying with spinning disc equipment has many advantages.

## REFERENCES

BIGGIN, P. (1979). Herbicides for use in forest nurseries, *Scottish Forestry*, **33** (1), 9–14

BIGGIN, P. (1980). Weed control in conifer transplant lines, *Proceedings Weed Control in Forestry Conference, 1980*, pp. 175–182

DALE, J. E. (1979). A non-mechanical system of herbicide application with a rope wick, *PANS*, **25** (4) 431–436

DAVISON, J. G. and BAILEY, J. A. (1980). The effect of weeds on the growth of a range of nursery stock species planted as liners and grown for two seasons, *Proceedings Weed Control in Forestry Conference, 1980*, pp. 13–20

FRYER, J. D. and MAKEPEACE, R. J. (eds.) (1977). *Weed Control Handbook*, vol. I, *Principles*, Blackwell, Oxford, 510 pp.

FRYER, J. D. and MAKEPEACE, R. J. (eds.) (1978). *Weed Control Handbook*, vol. II, *Recommendations*, Blackwell, Oxford, 532 pp.

HELLIWELL, D. R. and HARRISON, A. F. (1979). Effects of light and weed competition on the growth of seedlings of four tree species on a range of soils, *Q. J. Forestry*, **73** (3), 160–171

JOHNSON, M. S. and BRADSHAW, A. D. (1979). Ecological principles for the restoration of disturbed and degraded land, *Applied Biology IV* (ed. T. H. Coaker), Academic Press, London, 285 pp.

LUTMAN, P. J. W. (1980). A review of techniques that utilise height differences between crops and weeds to achieve selectivity, *B.C.P.C. Symposium on Spraying Systems for the 1980's*, 291–297

MAFF (1978). Narcissus bulb production, Booklet B2150, Ministry of Agriculture, Fisheries and Food, London, 79 pp.

MAFF (1979). *Review of the Safety for Use in the U.K. of the Herbicide 2,4,5-T.* Report of the Advisory Committee on Pesticides, Ministry of Agriculture, Fisheries and Food, London, 14 pp.

NEEL, P. L. and WILL, A. A. (1978). *Grevillea chrysodendron*: potential weed in South Florida, *Hort. Science*, **13** (1), 18–20

PARFITT, R. I., STINCHCOMBE, G. R. and STOTT, K. G. (1980). The establishment and growth of windbreak trees in polyethylene mulch, straw mulch and herbicide maintained soil, *Proceedings 1980 British Crop Protection Conference – Weeds*, pp. 739–745

ROBINSON, D. W. (1975/76). Herbicides in the landscape and garden (parts I, II and III), *Garden*, **100** (11), 554–559; **100** (12), 600–606; **101** (1), 35–41

ROBINSON, D. W. and KELLY, J. C. (1976). Further information on the tolerance of woody ornamentals to simazine. *Proceedings 1976 British Crop Protection Conference – Weeds*, pp. 333–339

ROBINSON, J. D. (1980). *Rhododendron ponticum* a weed of woodlands and forest plantations seriously affecting management, *Proceedings Weed Control in Forestry Conference, 1980*, pp. 89–96

SHEARING, S. J. and BATCH, J. J. (1979). PP 333– Field trials to control growth of amenity grasses, *Plant Growth Retardents*, Monograph 4 of the British Plant Growth Regulator Group, pp. 87–97

TAIT, F., BIGGIN, P. and McCAVISH, W. J. (1980). A review of trials with diphenamid in forest nurseries in Great Britain, *Proceedings Weed Control in Forestry Conference, 1980*, pp. 159–166

WATERHOUSE, D. P. (1980). Growth regulators in sward management, *Proceedings of Conference on Weed Control in Amenity Plantings*, University of Bath.

WELLS, T. C. E. (1980). *Creating Attractive Grassland using Native Species*, Nature Conservancy, London.

## FURTHER READING

HARDCASTLE, J. E. Y. (ed.) (1980). *Chemical Weed Control in Your Garden*, Grower Books, London, p. 24

PROCEEDINGS OF *Conference on Weed Control in Amenity Plantings*, University of Bath, 1980, compiled by R. J. Stephens and P. R. Thoday, p. 88

RORISON, I. H. and HUNT, R. (1980). *Amenity Grassland: an Ecological Perspective*, Wiley, Chichester, p. 272

# 12 FUTURE DEVELOPMENTS IN WEED CONTROL

## 12.1 BACKGROUND

Controlling weeds must remain of primary concern to man as it has been since the very first crops were grown. The pressure on farmers and growers to grow more food will intensify as the world population reaches six thousand million by the end of this century. Not only is world agriculture failing to feed the present population adequately but the land available for crops is being steadily reduced by building of all kinds, by erosion and by excessive salt accumulation in semi-arid lands. The sharp rise in oil prices that began in 1973 increased the costs of all material inputs to farming and especially those of fuel and nitrogen fertiliser; the realisation of high yields on the existing land has thereby been made yet harder.

The resulting economic pressure has increased the incentive to reduce the present enormous waste due to pests, diseases and weeds. The proportion of the potential crop that is lost is larger in those countries with less developed agriculture (Chapter 1) but changes introduced too rapidly can have serious social consequences, such as occurred when high yielding wheat and rice cultivars were introduced quickly and on a massive scale (Pearse, 1980). Although mechanisation and herbicides can reduce the drudgery of peasant agriculture any changes must be introduced sufficiently slowly to avoid destroying the social fabric of rural life and merely adding to the unemployed seeking their fortune in already over-crowded cities.

Writing of Southeast Asia, Mercado (1979) stated 'Most farmers use chemical weed control as the last resort or when they really become desperate.' By contrast, adoption of advanced technology in agriculture has been rapid in advanced countries. For example, the use of herbicides, which were seldom applied before 1945, has increased rapidly and now they are applied at least once per year to most cultivated land in Britain and North America. Generally the use of herbicides has increased more rapidly than that of other pesticides.

The present trends in weed control technology and predictions for future changes are considered below under three headings – herbicides, application techniques, and agronomy.

## 12.2 HERBICIDE DEVELOPMENTS

An increasing awareness of the potential hazards arising from widespread use of pesticides has led the agricultural authorities of most advanced countries to impose stringent regulations on their sale and use. Enforcement of the regulations in Britain is the responsibility of the Pesticides Safety Precautions Scheme (MAFF, 1979*a*) (Chapter 6). Manufacturers who wish to sell a new product are required to furnish detailed information about its toxicity to people, animals, wild plants and microorganisms. The maximum residues to be expected in crops, water and the soil and the likely accumulation of the active ingredient and its breakdown products in food chains must all be estimated. An independent committee of specialists (MAFF, 1979*b*) (Chapter 6) assesses the safety of the new compound, and reassesses the situation in the light of new information. It is this committee that has so far failed to find sufficient evidence to justify the general vilification of 2,4,5-T that has occurred in the media. A balanced view of risk and benefit arising from the use of new chemicals and methods must be adopted if technological innovation is not to be stifled; an independent enquiry into pollution arising from the use of chemicals in agriculture found that the present balance in Britain was right (Kornberg, 1979), but in some countries control over the use of toxic chemicals remains rudimentary (Mercado, 1979).

The search for new pesticides must now be concentrated on the pest problems of a few major world crops because of the cost of development, now in excess of ten million pounds (Braunholtz, 1977); therefore, horticultural and minor plantation crops must generally utilise chemicals that were developed for another purpose. For example, some of the herbicides currently recommended for red beet and strawberries were initially intended for use in sugar beet. There is a clear trend towards extending the use of existing well-tried products by using mixtures, different formulations, additives, chemical antidotes and novel application methods (Chapters 6 and 7). For example, in open ground nursery stock production simazine has advantages in cost and effectiveness over other alternatives but it can be toxic to some species. Advantage can be taken of its wide spectrum of control and relatively long persistence by applying a reduced and therefore safer, rate, while boosting its weedkilling effectiveness by addition of a reduced rate of another herbicide such as lenacil, oxadiazon or propyzamide. It has in the past been difficult for advisers to recommend the use of 'tank mixes' between the pesticides produced by different makers but some formal agreements between manufacturers for 'tank mixes' for agricultural crops have now been made (Trow-Smith, 1980).

There are still few working hypotheses from which the activity of new groups of herbicides could have been predicted and discoveries are made only after screening thousands of candidate compounds. Reliable prediction appears to be equally uncertain in the field of mixtures, synergistic additives and antidotes (Chapter 6).

Herbicides are usually differentiated from plant growth regulators, but it is impossible to draw too clear a line between the two groups. For example, maleic hydrazide (a growth regulator) and ethofumesate (a herbicide) can both be used to modify the vigour and composition of a grass sward, although their effects are quite different (Chapter 11). In addition, many herbicides have useful plant growth regulating properties; glyphosate and asulam both act as ripening agents in sugar-cane (Chapter 10). But the possibilities of using growth regulators in combination with herbicides has as yet hardly been explored; quite small changes in the relative vigour of crop and weed seedlings could have large effects on the competitive ability of the crop, and small changes in leaf surface or leaf orientation could markedly influence herbicide selectivity. The use of chemicals such as ethephon to stimulate weed seed germination has not proved to be either reliable or cheap enough for practical use, but sufficiently active compounds may yet be found. Unfortunately, although many plant growth regulators with potentially useful properties are known, they will not be marketed unless a use can be found for them in at least one major crop, because of the high cost of development.

## 12.3 DEVELOPMENTS IN APPLICATION TECHNOLOGY

Spraying machines much as we know them today were first devised in the late nineteenth century in order to apply fungicides to vines and insecticides to control Colorado beetle and other pests. Although the equipment is now lighter, corrosion resistant and more efficient, it relies upon the same underlying principles of operation. Most agricultural chemicals are still applied by forcing liquid under pressure through nozzles to create a spray (Chapter 7), and the definition of spraying given in 1886 (quoted by Heijne, 1980) as 'the throwing upon plants of any fluid or semi-fluid in the form of fine rain or mist' shows that methods have changed little since then. However, by using improved equipment and with greater knowledge of the chemical and physical interactions between sprays and sprayed surfaces (Chapter 6), application has been refined and the volume of liquid required and sometimes the amount of active ingredient also, have been reduced. Better control of pests, at less cost and with reduced contamination of the environment, including unwanted harm to pest parasites and predators is now possible. Hydraulic sprayers have been much improved by using newer materials, especially plastics, and major innovations have included spinning disc sprayers and electrostatic sprayers such as the 'Electrodyn'; these machines for the first time make it possible to avoid creating the small satellite drops that are inevitably formed when liquid forced through nozzles under pressure breaks up, first into ligaments and then into drops. Deliberate use of drifting droplets less than 150–100$\mu$m diameter has made possible the control of some insect pests at volume rates of around 1.0 l/ha, a technique known as ultra low volume (ULV) spraying. When larger droplets of 200–300$\mu$m are produced by a spinning disc machine such as the Micron 'Herbi' a higher volume rate of about 20 l/ha is probably required to achieve adequate coverage for foliar-acting herbicides to be effective. To distinguish between application with larger droplets of reasonably uniform size and ULV spraying the term controlled droplet application (CDA) was devised to cover situations in which a spray of non-drifting droplets of uniform size was applied.

One of the benefits of traditional hydraulic spraying is that the relatively wide spectrum of droplet size produced from nozzles (Chapter 7) enables an acceptable control of pests to be achieved under widely different circumstances. The ability to produce a spray with a very narrow size spectrum does not guarantee that the spray will be more effective, although the benefits of eliminating both very small drifting drops that can damage nearby crops and contaminate the environment, and the extra large ones that bounce and splash away from the target, seem obvious at first sight. Retention on plant surfaces is often greater with drops below 150$\mu$m (Chapter 7); however, the drift problem with herbicides generally precludes the use of such small droplets, an exception being asulam which can be drifted over extensive upland pasture to control bracken. Were it possible to use smaller drops for all herbicide spraying, yet avoid the dangers of drift, then a reduction in volume of spray and perhaps also of active ingredient should be possible.

In fruit tree spraying the arrival on target of small driftable drops can be enhanced by carrying them in a moving air stream, but such small drops will be largely lost in conventional herbicide ground spraying. The idea of giving an electrostatic charge to small drops, thereby attracting them to the target which carries an opposite charge, is a fascinating one but until recently no practicable way of achieving this in the field had been devised. One recently introduced method is to charge the droplets as they leave the nozzle, using a moving air stream to carry the cloud of charged particles to and within the crop canopy, relying on the electrostatic charge to achieve deposition on the target; machines working on this principle are claimed to be capable of halving the required dose of insecticide (Law, 1980). Another method now being tried is to charge the drops produced by a spinning disc sprayer (Arnold and Pye, 1980). An entirely novel approach has been adopted in the 'Electrodyn' sprayer, which relies on neither gravity nor wind to convey the drops to the target, but instead almost wholly on the electrostatic field. Capable of producing a narrow spectrum of droplets, which can be accurately controlled between 40 and 200$\mu$m diameter by varying the input voltage, drops are produced by 'applying a coulombic field force directly to the surface of the liquid, thus setting up standing waves from each crest of which issues a uniform jet of charged liquid' (Coffee, 1980); the drops from the 'Electrodyn' are not only charged but propelled to the target without the need for moving parts or nozzles. The results of trials on many crops with several types of pesticide show that the recovery on the target of applied spray from the 'Electrodyn' is always better than from other types of sprayer, especially when the drop size is small enough for drift. Small batteries provide the power for prolonged use but a disadvantage of the system is that the requirement for spray liquids to have rather precise physical characteristics will create formulation problems.

There is an increasing interest in application of herbicide granules, especially for small or irregularly shaped areas and for spot treatment of perennial weeds. Fungicides and insecticides are used most economically in seed dressings and there is a possibility that herbicides could also be applied in this way; already in North America lucerne (= alfalfa) has germinated unharmed into a weed-free strip about 6 cm wide when the seed sown had been previously coated with EPTC. This method saves herbicide, reduces environmental contamination and makes a separate application operation unnecessary.

The new methods of application will be adopted for situations where a benefit can be clearly shown. For example, spinning discs and the 'Electrodyn' should find general acceptance for small scale application, especially in peasant agriculture. But hydraulic sprayers will continue to be used for most herbicide application to large scale crops for many years to come. These machines have benefited from advances in technology and materials science. Use of strong plastics and stainless steel for tanks, pumps and other parts has reduced weight and corrosion; effective filtration and agitation reduces blockages and uneven application. Microprocessors are now used in the larger machines, providing the operator with a display showing the performance of the sprayer derived from major functions such as flow rate at each nozzle and forward speed. Electronic devices can be mounted on booms to show the operator how well he is matching his spray with the previous bout. The cheap, versatile and small electronic equipment now available provides unprecedented opportunities for the innovator (Allan, 1980; Lawrence, 1980).

## 12.4 CHANGES IN CROP AGRONOMY

Herbicides have facilitated a revolution in the crop agronomy of developed countries. Rotation and cultivation, formerly essential for weed control, have been progressively reduced or abandoned (Chapter 8). Some crops can now be sown, grown and harvested entirely by machine, with a considerable saving in labour requirements. Minimum cultivation methods are increasingly being used in the production of cereal and some other farm crops in Britain, but in North America maize, cotton, peanuts, tobacco and sugar beet have also been grown successfully without ploughing (Day, 1978). Not only do agricultural crops now require a minimum of human effort to be expended, but formerly labour intensive crops such as strawberries and vegetables are now grown extensively instead of in small lots near to a supply of labour.

During periods of economic growth those people leaving the land permanently for the city can expect to find useful employment but the depopulation of rural areas creates many social problems. While the use of herbicides merely as a substitute for labour in developing countries without expanding industry leads to unemployment, their judicious use to control water and parasitic weeds and poisonous plants must improve the lot of the rural community. However, the present trend is for development in weed control technology to be adopted most rapidly in the developed world, and this seems likely to continue.

Problems have arisen when new techniques that work well on a small scale are attempted on larger areas. Early experience with direct drilling of cereals in Britain showed an overall reduction in yield of about ten per cent but as the design of drills was improved, slug damage to seedlings reduced and farmers became more skilled, the yields improved and by 1977 direct drilled crops were generally as good if not better than those cultivated in the normal way (Ellis *et al.*, 1977). Problems associated with changed methods of tillage and weed control in perennial crops may take longer to reveal themselves. Experience with fruit trees under total weed-free management with herbicides, for example, suggests that the slow downward movement of bases over several years may so deplete the soil in the rooting zone of available calcium that storage breakdown of the fruit becomes more prevalent. With hindsight it now seems obvious that sufficient calcium to allow for loss by leaching should be incorporated at planting time. Some problems arise directly from the elimination of cultivation, such as the proliferation of unwanted suckers in plums and cherries and in raspberries. However, such problems have so far proved to be minor ones when compared with the advantages that the elimination of cultivations has conferred in many crops.

The adoption of herbicides as an alternative to cultivation for weed control in direct drilling and other minimum tillage systems has led to a substantial saving in fuel requirement (Chapter 8), although the greatest overall savings in energy have sometimes been associated with a combined use of herbicides and limited pre-sowing tillage, rather than with the elimination of all cultivations (Patterson *et al.*, 1980). The fuel saving from these systems is worthwhile but is small compared with the large energy requirements for the production of nitrogen fertiliser and in food processing. Fuel for spraying can be saved by reducing the volume applied and by using lighter, faster spraying machines (Elliott, 1980); spraying can also be combined with drilling or other necessary operations to save fuel and time.

Chemical weed control has produced profound quantitative and qualitative changes in the weed flora of farmland. A reduction in the numbers and variety of wild life, often ascribed to pesticides, is probably due more to a reduction in the food when weeds are adequately controlled than to direct effects of the pesticide; certainly this is true for herbicides (Chapter 6). A good case can be made for the retention of wild life sanctuaries in the form of hedges and woods both for aesthetic reasons and as a source of parasites and predators of pests.

Interspecific selection between weeds has been one inevitable result of using herbicides. Grass weeds, for example, have increased as other weeds have been controlled but selection between broad-leaved species also occurs. Mayweeds proliferate in lettuce crops when herbicides ineffective against members of the Compositae are applied, and in cereal crops sprayed with 2,4-D and MCPA a series of weeds became dominant over a period of years, each to be controlled in turn as new herbicides were discovered (Chapter 6). Selection of resistant forms within species, as occurs all too frequently in fungal pathogens and insect pests, has occurred in a few weed species after prolonged and repeated application of triazine herbicides but the problem has been easily overcome by 'ringing the changes' with herbicides other than triazines. There is a real possibility that the same occurrence of resistance genes could be selected for in crop plants, as has already been achieved in paraquat-resistant perennial rye grass (Faulkner, 1976).

Of greater potential concern than selection within the existing weed flora is the introduction of new weeds, or the creation of weed forms such as weed beet, which renders conventional use of selective herbicides impossible (Parker, 1977; Hornsey and Arnold, 1979) (Chapter 2). Such problems are often impossible to eliminate by the time they have been recognised, and containment is then all that can be expected. The noxious nature of the weed may be more fully expressed in the new environment, but unless a new introduction is first tested in isolation there is no way of confirming this in advance.

Agricultural scientists are often exhorted by the lay public to perfect biological control as an alternative to pesticides. There have been notable successes in the control of weeds (Chapter 5) but scope for future action is limited. The activity of biocontrol agents must be highly specific – if this were not so they would be equally damaging to crops. It follows that only individual weed species sufficiently unlike the local crops are likely candidates for the introduction of self-perpetuating biocontrol agents. For example, before the rust *Puccinia chondrillina* was released in Australia to control Skeleton Weed (*Chondrilla juncea*), its inability to colonise related crops such as lettuce had first to be established. Work on the biological control of major weeds including Water Hyacinth (*Eichhornia crassipes*), Lantana (*Lantana camara*) and the nut-forming sedges (*Cyperus rotundus* and *C. esculentus*) is being actively pursued, but it seems likely that the control achieved will generally augment other methods rather than replace them.

## 12.5 WEED CONTROL IN THE FUTURE

An attempt has been made to predict some of the changes that seem likely if current trends continue; however, predictions of the future are notoriously unreliable. The weed scientist will be expected to solve existing problems effectively and cheaply and to cope with new ones as they arise. Weed science presents a great diversity of scientific challenges, as demonstrated by an increasing output of research results published in journals such as *Weed Research* in Europe and in *Weed Science* in North America. It once seemed that new 'wonder' herbicides with remarkable properties were being announced so frequently that all the world's weed problems would ultimately be cured, permanently. But weeds have proved to be resilient and adaptable and it is now realised that weed 'control', that is keeping the level of weediness sufficiently low for our immediate purpose, is the most that we can expect; with few exceptions, and then only on a local scale, eradication is seldom a realistic aim. In view of the need for protective plant cover to prevent soil erosion it is better that the ultimate weapon of total weed suppression is not available to us. Instead, we must learn to live with weeds.

## REFERENCES

ALLAN, J. R. McB. (1980). Developments in monitoring and control systems for greater accuracy in spray application, *BCPC Symposium on Spraying Systems for the 1980's*, pp. 201–213

ARNOLD, A. J. and PYE, B. J. (1980). Spray application with charged rotary atomisers, *BCPC Monograph 24, Spraying Systems for the 1980's*, 109–118

BRAUNHOLTZ, J. T. (1977). The crop protection industry; products in prospect, *Proceedings 1977 British Crop Protection Conference – Pests and Diseases*, pp. 659–670

COFFEE, R. A. (1980). Electrodynamic spraying, *BCPC Monograph 24, Spraying Systems for the 1980's*, 95–108

DAY, B. E. (1978). The status and future of chemical weed control, in *Pest Control Strategies* (eds. E. H. Smith and D. Pimental), Academic Press, New York, 334 pp.

ELLIOTT, J. G. (1980). Low volume, low drift and high speed – a great new opportunity, *BCPC Symposium on Spraying Systems for the 1980's*, pp. 175–183

ELLIS, F. B., CHRISTIAN, D. G., GRAHAM, J. P. and JACKSON, R. (1977). Long-term tillage experiments, *A.R.C. Letcombe Laboratory, Annual Report*, p. 31

FAULKNER, J. S. (1976). A paraquat resistant variety of *Lolium perenne* under field conditions, *Proceedings 13th British Weed Control Conference*, pp. 485–490

HEIJNE, C. G. (1980). A review of pesticide application systems, *BCPC Monograph 24, Spraying Systems for the 1980's*, pp. 75–83

HORNSEY, K. G. and ARNOLD, M. H. (1979). The origins of weed beet, *Ann. Appl. Biol.*, **92**, 279–285

KORNBERG, H. (1979). *Royal Commission on Environmental Pollution – 7th Report – Agriculture and Pollution*, HMSO, London, p. 280

LAW, S. E. (1980). Droplet charging and electrostatic deposition of pesticide sprays – research and development in the U.S.A., *BCPC Monograph 24, Spraying Systems for the 1980's*, 85–94

LAWRENCE, D. C. (1980). Aids to swath matching in tractor spraying, *BCPC Symposium on Spraying Systems for the 1980's*, pp. 215–221

MAFF, (1979*a*). *United Kingdom Pesticides Safety Precautions Scheme*, revised 1979, Ministry of Agriculture, Fisheries and Food, London.

MAFF (1979*b*). *Review of the Safety for use in the U.K. of the herbicide 2,4,5-T*, Report of the Advisory Committee on Pesticides, Ministry of Agriculture, Fisheries and Food, London.

MERCADO, Beatriz L. (1979). *Introduction to Weed Science*, Southeast Asian Regional Center for Graduate Study and Research in Agriculture, Laguna, Philippines, p. 292

PARKER, C. (1977). Prediction of new weed problems, especially in the developing world, in *Origins of Pest, Parasite, Disease and Weed Problems* (eds. C. H. Cherrett and G. R. Sagar), British Ecological Society 18th Symposium 1976, pp. 249–264

PATTERSON, D. E., CHAMEN, W. C. T. and RICHARDSON, C. D. (1980). Long-term experiments with tillage systems to improve the economy of cultivations for cereals, *J. agric. Engng. Res.*, **25**, 1–35

PEARSE, A. (1980). *Seeds of Plenty, Seeds of Want*, Clarendon Press, Oxford, p. 262

TROW-SMITH, R. (1980). Tank mixes, *Farmers Weekly (Supplement)*, Feb. 1980, 62 pp.

# Index